ES&T Presents

TV

Troubleshooting
& Repair

PROMPT®

PUBLICATIONS

A Division of
Howard W. Sams & Company
Indianapolis, IN
A Bell Atlantic Company

International Standard Book Number: 0-7906-1086-8

Library of Congress Catalog Card Number: 96-70019

Acquisitions Editor: Candace M. Drake
Editors: Natalie F. Harris, Nils Conrad Persson
Contributing Editors: Steve Babbert, Homer L. Davidson, Hulon Forrester,
        Ricky Hall, Brian E. Jackson, Glen Kropuenske, Conrad Persson,
        R.D. Redden, Lamarr Ritchie, Dale Shackelford, and the ES&T Staff
Typesetter: Candace Drake, Natalie Harris
Cover Design: Suzanne Lincoln

Illustrations & Text: Supplied by *Electronic Servicing & Technology* (*ES&T*) Magazine, CQ Communications, Inc., 76 N. Broadyway, Hicksville, NY 11801.

Table of Contents

Chapter Ten
Electronic Tuner Theory and Troubleshooting

Preface

TV set servicing has never been easy. The operation of electronics circuits is something of a mystery even to those whose profession it is to design, produce, test or service electronics. Unlike mechanical systems, whose operation may be observed directly, the operation of electronics circuits can only be observed via sophisticated instruments and the imagination of the practitioner.

To further complicate matters, electronics circuits and devices are constantly evolving. In fact, this change is so rapid that it might be better considered as a revolution. Components are constantly becoming smaller, more complex, and frequently more specialized in purpose.

Another complicating factor is that even though TVs are constantly becoming more sophisticated and are producing better pictures and sound, the price of the products stays the same, or decreases. Frequently, therefore, it seems to make more sense to a consumer to discard a TV than to spend the money to have it repaired.

All of this makes it increasingly difficult for a consumer electronics service center to remain in business. The service manager and service technician need timely, insightful information in order to be able to locate the correct service literature, make a quick diagnosis, obtain the correct replacement component(s), complete the repair, and get the TV back to the consumer.

The articles in this book, originally published in *Electronic Servicing & Technology* magazine, were written by professional technicians, most of whom service TV sets every day, and must be good at it in order to prosper. These articles provide general descriptions of television circuit operation, detailed service procedures, diagnostic hints and tips, and more.

The information presented here will make it possible for technicians to service TVs faster, more efficiently, and more economically, thus making it more likely that customers will choose not to discard their faulty products, but to have them restored to service by a trained, competent professional.

Chapter One

Is It the CRT That's Bad?
By Hulon Forrester

The cathode ray tube (CRT), or picture tube, has been an integral part of our everyday living for so long that it is just as commonplace as the steering wheel on a car. Everyone knows it's there, but no one thinks very much about it until it doesn't work.

In the many years that I have been in the business, I can't remember seeing any article that describes how a tube works, how to determine when there is a problem with a CRT in a TV set, or whether the problem is in the tube or in the circuits supplying it.

The Importance of the CRT

In the almost one-hundred year history of the CRT, the last fifty years have seen the cathode ray tube grow into one of the most important parts of electronic equipment, yet it has been taken for granted as a black box. Some of the most recent and bestselling publications describing how CRTs work contain errors. Even manufacturers of CRT rejuvenators are not always totally correct in describing exactly what their equipment does.

TV service technicians have lived with TV tube replacement long enough to feel at home with it, but there are many computer technicians who have not had the opportunity to learn as much as they would like to. Many trade and technical schools avoid servicing that has to do with the CRT. As an example of the ignorance that exists about CRTs, a customer called me not too long ago and said, "The picture tube I bought is no good. It has a straight line across the center of the screen!" Of course, this is an indication of a problem in the vertical deflection circuits, not in the tube.

Many Types of CRTs

Over the years, more than 5,000 different types of cathode ray tubes have been developed for many high volume applications. Thousands more have been produced for special applications. All of this arises out of the development of RADAR during World War II, from which evolved the high-frequency circuits necessary for the introduction of TV immediately after the end of the war in 1945.

The Continuing Demand for CRTs

The instant popularity of television created a never ending demand for larger and larger screens that continues to this day. Color, which was introduced in 1957, created a larger TV audience than anyone could have ever imagined. The tremendous growth of the data display market, including medical, aircraft, automotive, navigational and computer electronics, caused an explosive demand for all types of displays that produced images in monochrome and color.

It's Important to Understand CRTs

The cathode ray tube is still superior in reliability, cost, and performance to anything else in use or proposed for use. Flat LCD panels are great for portable equipment, watches and other applications where small size and light weight are essential, but when it comes to performance at the lowest cost, where space, weight, and power are of lesser consequence, the CRT is superior. The CRT is so well entrenched that it will be in use for many years to come even if a better substitute were introduced tomorrow.

It is important for those of us in the business to understand CRTs in order for us to get the most out of them, as well as to properly replace them when necessary. With HDTV, DATV and high resolution data display already here or on the horizon, understanding CRTs is even more important for proper maintenance. Today, there are enough tubes in service that each human being in the world could have a screen to themselves.

Two Types of Monochrome Tubes

There are two basic monochrome tube types: those that employ electro-static deflection, and those that employ magnetic deflection. Electro-static deflection tubes have deflection plates built inside the tubes. (*Figure 1-1.*) The more commonly used magnetic deflection tubes require a yoke mounted on the neck of the tube to deflect the electron beam. (*Figure 1-2.*)

Electrostatic deflection tubes are still the only way to look at waveforms in "real time." You will find this tube used in oscilloscopes and some types of medical equipment. Most electrostatic tube screens are no more than seven inches in diameter.

Magnetic deflection tubes may have screens smaller than one inch and larger than 35 inches across. These require a magnetic deflection yoke on the outside which replaces the deflection plates on the inside of the elec-trostatic types.

Although electrostatic tubes for special applications have been built in larger diameters, modern TV and computer monitors all use magnetic deflection.

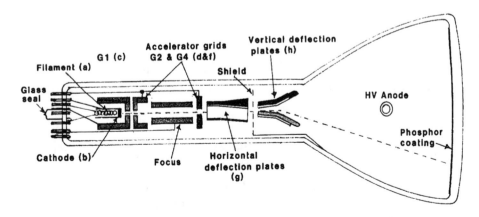

Figure 1-1. The deflection plates in electrostatic deflection tubes are built inside the tube.

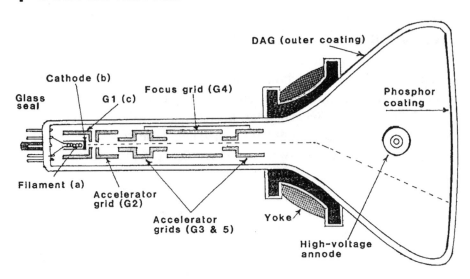

Figure 1-2. The more commonly used magnetic deflection tubes require a yoke mounted on the neck of the tube to deflect the electron beam.

Electrostatic Deflection

Electrostatic deflection tubes (*Figure 1-1*) are superior when a wide variety of frequencies need to be viewed, such as in an oscilloscope. The better CRTs enable the user to view frequencies from DC to several gigahertz. No other display method has this capability.

In electrostatic deflection, the signal to be viewed is fed to the vertical (Y) plates. The output of an internal variable linear sweep circuit is fed to the horizontal (X) plates. The signal produced by this sweep circuit is at the same frequency as the signal that is to be observed, or a fraction of that frequency.

For example, in order to view a single cycle of a 60 Hz waveform, the frequency of the horizontal waveform must be a linear 60 Hz. This waveform is called a sawtooth waveform (*Figure 1-3*) because of its appearance, but it must move the electron beam in the tube horizontally at a constant rate of speed in order to faithfully reproduce the signal applied to the vertical plates.

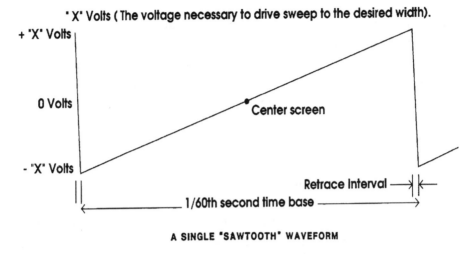

A SINGLE "SAWTOOTH" WAVEFORM

Figure 1-3. *The waveform that provides horizontal sweep in a CRT is called a sawtooth waveform because of its appearance. It must move the electron beam in the tube horizontally at a constant rate of speed in order to faithfully reproduce the signal applied to the vertical plates.*

The vertical line of the sawtooth is not totally vertical. It represents the time required for the beam to return to the left side of the screen to begin subsequent linear sweeps. This is known as the return trace or flyback time. This quick pulse is used in magnetic deflection circuits to generate high voltage and is where the term flyback transformer comes from.

All horizontal sweep waveforms must be linear. If these waveforms are not linear, the picture is distorted. This applies to a raster or a single line sweep in either the electrostatic or magnetic deflection format. In essence, it is a time base. Any half inch of that line represents the same time period in any other half inch.

Displaying Multiple Cycles of a Waveform

In order to display more than one 60 Hz image of a waveform on the oscilloscope screen, the frequency of the horizontal waveform must be some fraction of the vertical frequency. For example, in order to see four cycles of a 60 Hz waveform, the horizontal frequency must equal the vertical frequency divided by four, or 15 Hz

During a single horizontal sweep of 15 Hz, four 60 Hz images are fed to the vertical plates. This results in four 60 Hz images being shown on the screen. The last reproduction of the last waveform will not quite reach the base line because the retrace time takes that last fraction of a second to return to the beginning of the next sweep.

Theory of Operation of a CRT

In *Figure 1-1*, the filament heats the cathode which emits electrons. The brightness of the beam is determined by the degree of negative voltage on the grid (G1) relative to the cathode. If the voltage on G1 is sufficiently negative, no electrons will pass and nothing will show on the screen. Because of the effect of the grid on the electron beam, the brightness control usually controls the voltage of G1. The exact point where electron flow stops is known as the cutoff voltage.

The next element after G1 is G2. G2 and G4 are usually tied together, and the focus grid (G3) is sandwiched in between. Grids G2 and G4 are accelerator grids, and G3 is the focus anode which "puts a point" on the electron beam.

The screen is at a potential of greater than 2,000V with respect to the cathode, which attracts the beam. The electron's last obstacle to the screen is the deflection plates: two vertical and two horizontal. Depending on the polarity, voltages applied to those plates will deflect the beam up or down (vertical) and side to side (horizontal).

Deflection Angle

The X and Y plates are in a "V" configuration which allows the beam to bend vertically and horizontally without touching either set of plates. The maximum angle at which the plates allow the beam to bend in a combined vertical and horizontal configuration is called the deflection angle.

You will notice that the deflection angle in an electrostatic tube is very low; some 50 degrees or less. A large tube in this configuration would be quite long. In fact, a 21-inch-diagonal tube would be over three feet long, which would make this tube type somewhat cumbersome in manufacturing cabinets for large screen units.

Manufacturing an inexpensive source of high voltage for these tube types is not easy. Usually, the voltage difference between the cathode and the screen is around 4,000V. This is achieved by a more complex version of the voltage doubler circuit which can get about 2,000V from the 120V AC power. By creating a -2,000V supply to the cathode and a +2,000V supply to the screen, the 4,000V difference is achieved. Therefore, it is not unusual to find the cathode, G1, G2, G3 and G4 "floating" at about a -2,000V below ground potential.

Electromagnetic Deflection

The electromagnetic deflection scheme is where the yoke comes into being. First, the yoke (*Figure 1-2*) replaces the electrostatic deflection plates inside the tube. In a rectangular tube, it is difficult to align the electron gun with the sides of the tube so that a rectangular picture would be in line within the rectangular screen. This alignment is done by adding an extra winding around the neck of the tube with a variable low voltage applied to it. This is called a twist coil.

With a yoke, it is simple to position the picture within the tube by twisting the yoke. Also, since the yoke is outside the tube, there is no danger of the electron beam striking the deflection plates and the coil design permits a higher deflection angle; 114 degrees is common. This allows the gun to be closer to the screen, which in turn allows the cabinet for even a 25-inch screen to sit close to the wall.

Electromagnetic Sweep Circuits

Sweep circuits may now be applied to the yoke while focus, brightness and the video signal are applied to the tube. In addition, the neck may be smaller, which in turn requires less power for the sweep circuits. (The closer the magnetic field is to the electron beam, the less power is required to deflect it.)

The absence of internal deflection plates also allows for a shorter neck overall. You may note that the vertical sweep circuit may fail, causing a single horizontal line to appear across the tube. A vertical line on the tube can only be caused by an open horizontal yoke winding. The high voltage supply and the flyback transformer is dependent on the horizontal sweep circuit working properly. If the sweep fails, the HV goes too, causing a blank screen instead of a vertical line.

A short in a single turn of either yoke winding will cause the raster to turn into a keystone pattern: the top of the raster may be wider than the bottom or vice versa; or one side higher than the other, depending on the location of the short.

If you should encounter a set in which the display on the CRT screen is a vertical or horizontal line, use a probe to test the solder joints of the yoke windings. Sometimes the solder flux will cause the wire to deteriorate at the joint. It will appear to be conductive, but it isn't.

In a typical magnetic deflection tube, G1 is the control grid; G2, G3, and G5 are accelerator grids; with G4, the focus grid. The voltage on G2 is an indicator of resolution. A voltage of 100V is low resolution, and 300V and up indicates high resolution. In a typical circuit, the signal is fed to the cathode. G3 and G5 are connected to the screen and use the HV as an accelerator voltage, and G4 is the focus anode that is connected to the focus control.

In some modern tube designs, G4 is grounded, and there is no focus control. G1 is typically connected to the brightness control and the negative voltage necessary to cut off the electron beam is known as the cutoff voltage. You might note that inability to control the brightness is usually indicative of a cathode to grid short. There are always exceptions to circuit design. Some signals are fed to G1 rather than the cathode.

Data Display CRTs

Monochrome data display tubes come in a variety of colors like amber, green, page white, and black and white. You may replace a CRT of one screen color with one of another color, provided that the persistence is the same. Persistence is the length of time that the phosphor remains lit after the electron beam leaves it.

For example, P31 and P39 (the pacific rim uses "B" instead of "P" but otherwise it is the same) are the same color but of different persistence. Some IBM monochrome monitors use P39, and if the phosphor is replaced with any color that has a lower persistence, the picture flickers. So be careful to maintain the same persistence when you change color.

This mistake in persistence doesn't occur often when using the same color. The error usually happens when the user wants to go from a green screen to amber or page white.

Solving CRT-Related Problems

If the screen of a CRT is blank, in most cases you should look at the back of the tube and see if the filament is lit. Typical filament voltages are 6.3V to 12.6V. If you see a warm glow, you know that the filament is lit.

Electron beam flow starts at the cathode and in most cases, the video signal is fed to the cathode. There should be a resistor from the cathode to ground and a capacitor from the output transistor to the cathode. If there is no current flow from the cathode, there will be no voltage drop across the resistor. If the capacitor is shorted, the voltage will be the same on both sides of the capacitor. If you find nothing remarkable, turn the unit off and measure the resistance between the cathode and G1 with an ohmmeter on a low scale. (A hot cathode will give you a false reading.) With the unit on, if there is no short between the cathode and G1, you will also see a change in the voltage when you turn the brightness control up or down. On the other hand, a cathode to G1 short will cause a bright, usually out of focus picture, and the brightness control will have no effect. The tube is bad.

A gassy tube will give you a bright blue glow in the neck of the tube. If there is enough gas in the tube, it will short out the high voltage. It is a good idea at this point to see if you have high voltage. Usually, the "rushing" sound that a TV makes is the sound of the high voltage being applied. That sound is a clue that there is high voltage at the CRT. If that sound is not conclusive, you can test the high voltage with an high voltage meter.

Remember, the high voltage supplied to a monochrome CRT may be 20,000V or more. Always use extreme caution when measuring this voltage.

The picture on a weak gassy tube will "balloon" when you turn up the brightness and/or go negative. The tube is also bad. If there is air in the tube, you will see arcing inside the neck of the tube. High voltage will not arc in a vacuum.

You may choose any sequence you want to take in these steps; the order is not important as long as you cover all of the bases. Another two things to check are G2, which should be at least 100V and the focus voltage which varies from a minus voltage through zero to a positive voltage. If you see an extra lead coming from the flyback to the focus anode, the focus voltage could be several thousand volts, but would not normally kill the picture. Some G4s in newer types are connected to ground and have no focus control at all.

Summary of Monochrome Tubes

The most important thing to remember in checking a monochrome tube is that it should have filament voltage, high voltage, G2 voltage, and brightness control. If the cathode is shorted to G1, you can measure that with an ohmmeter. You will also find that the brightness control will not have any effect on the voltage of G1.

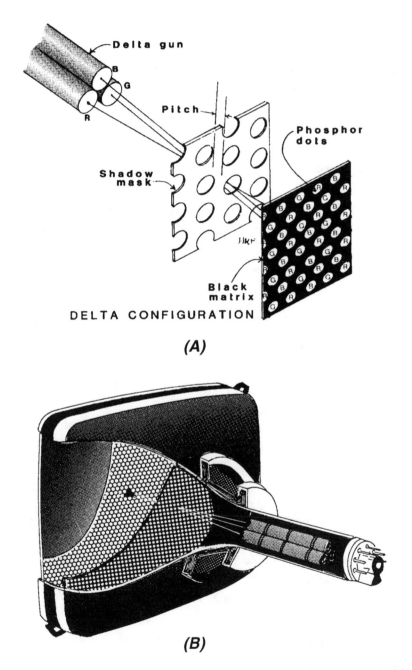

DELTA CONFIGURATION

(A)

(B)

Figure 1-4. *In a delta-gun CRT, the guns are in a delta (triangular) configuration. The drawing marked "A" is a detail of the mask, which prevents electrons from one color gun from striking phosphors of the other colors.*

If you have no high voltage, check the horizontal circuit and transistor. If you have a horizontal line, check the vertical circuit including the vertical transistors, output transformer and the vertical winding on the yoke. (Turn down the brightness so you won't burn the tube. A complete picture compacted into one horizontal line packs a hard punch and will easily burn the phosphor.)

Color Tubes

In the neck of a color tube there are three monochrome guns which fire three electron beams at the screen simultaneously except when the screen area is meant to be black. The guns may be in a delta (triangular) configuration, or side by side, also referred to as in-line (*Figures 1-4* and *1-5*) depending on the tube and mask design. *Figures 1-4A* and *1-5A* are simply engineering drawings to give more detail of the two systems. The mask is simply a gate which blocks each electron beam from striking phosphors of the other two colors.

Convergence

When the electron guns are adjusted into that configuration, it is called convergence. Convergence of the dot matrix configuration is more demanding than the stripes, but the principle is the same. The smaller the dots, the closer they are together. This measurement is called pitch.

The smaller the pitch, the higher the resolution; but a smaller pitch usually requires more horizontal lines and possibly a higher vertical frequency as well. Together they make a sharper picture. However, strange patterns called moire patterns appear with the image when a tube with the wrong pitch is used with improper sweep frequencies.

Refresh Rate

TV broadcast must use the standards, set by the FCC in 1945, of 525 horizontal lines and a refresh rate of 60 Hz, which is the NTSC standard.

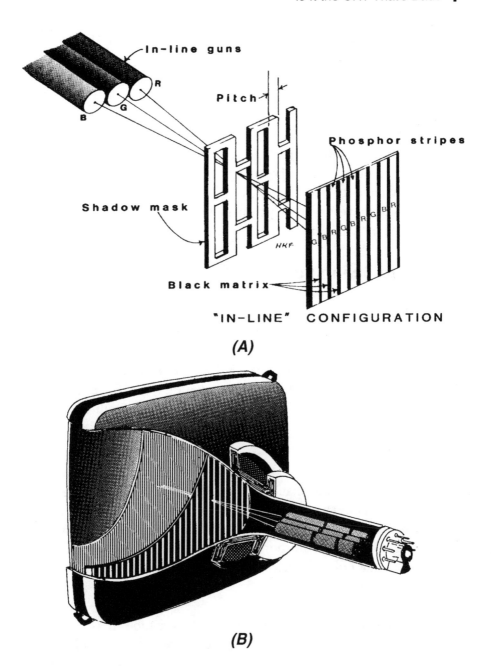

Figure 1-5. In another type of CRT, the guns are side by side, also called in line. The drawing marked "A" is a detail of the mask which blocks each electron beam from striking phosphors of the other two colors.

For a sharper picture, computer technology and new TV designs may have as many as 1050 lines in a picture. Some ultrahigh resolution (usually monochrome) CRTs have as many as 4,000 lines and approach 35mm film quality. The refresh rate may be higher also.

The minimum recommended refresh rate so that any flicker of the video image is invisible to the human eye is 72 Hz. CAD/CAM and other high resolution computers are not bound by the TV standards. Each complete system may have its own standards, so both the horizontal and vertical frequencies could very well vary from system to system. The basic theory, however, remains the same.

Interlaced Scanning

The old TV standard raster also produced an alternate line raster (interlaced scanning) which was designed to reduce flicker, but did affect the sharpness of the picture. In interlaced scanning, within the first 30th of a second, the odd lines were drawn (1,3,5,7,9, etc.) and within the second 30th of a second, the even lines were drawn (2,4,6,8,10 etc.). The new TV circuits and data display circuits will not have this feature.

No Picture

If there is no picture, the chances are that the high voltage is gone. If the tube is gassy, it could be bad enough to short out the high voltage and kill the picture. The high voltage should be the first thing to check. Since the high voltage comes from the horizontal sweep circuit, the cause could be anywhere in that circuit if disconnecting the high voltage lead does not bring back the high voltage.

In a color tube, in essence, you have three picture tubes in one envelope with a red, blue and green picture. If the color of a picture is bad, it may be in the color circuits in the set, or it could be that one gun is bad or shorted. If the screen is one color when the unit is first turned on and another color after it has been on a while, or if it is out of focus with a blue glow in the neck, the tube is weak or gassy or both. It is rare indeed when all three guns fail at the same time which makes it easier to tell whether the problem is the tube or the circuit.

Phosphor Burn

A computer monitor may sit for hours with a single image on the screen; this often causes the phosphor to "burn." If you begin to see the image on the screen after the monitor is turned off, the screen is burned. Prolonged use of the monitor after the burn begins to show when the monitor is off will permanently damage the tube and make it irreparable.

In many instances, a light burn in a color tube may be repaired if the tube is remanufactured. A recycled monochrome tube also has a new phosphor. However, a badly burned monochrome tube may have irreparably damaged glass.

Symptoms of Power Supply Problems

A double dark bar across the screen means that AC is getting through your power supply. Depending on where your sync voltages are coming from, the bars may roll or be stationary. If you will look at an AC power supply, one half of each cycle is turned into 120 Hz of pulsating DC, then smoothed out to smooth DC with chokes and filter capacitors.

A bad capacitor will allow a 120 Hz ripple. A single bar across the screen means a cathode to filament short in the CRT, because that is 60 Hz AC. The tube is bad, but sometimes an isolation transformer or a DC filament supply may prolong the use of the tube.

Color Display Variations

There are "hybrid" and some "off-the-wall" types of color displays. For example, there are some in-line guns used with a dot matrix and vice versa. These are most common in high resolution CRTs. Then there are projection tubes and some flat CRTs used in miniature portables where the electron beam curves to the screen from a gun parallel to it. Zenith has developed a flat screen CRT that has some unusual characteristics, but built on the same basic principles as a standard tube.

Refinements continue to be developed. No matter how good a CRT is, someone is always trying to make it better or larger. The sharpest possible picture is still from the original design, the dot matrix. Some years ago, General Electric used a 13-inch color tube and a lens to make a projection color set but it was not bright enough. Now GE is producing their Talaria projection system that is very complex, expensive, and used only in theaters. Sharp has come out with a system using a high intensity lamp and three liquid crystal screens converging through a single lens. This system will focus a picture from about 25 inches to 15 feet. It does not use a CRT at all.

Separate CRTs for Each Color

Two color video display systems use separate red, green and blue tubes. When the pictures from these three tubes are superimposed into a single picture, a color picture is produced. One is a tube called a Novabeam,

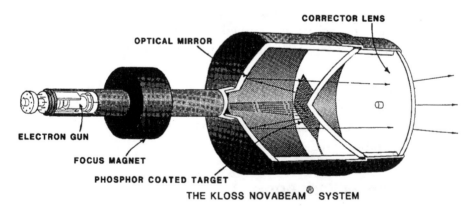

THE KLOSS NOVABEAM® SYSTEM

Figure 1-6. Some projection systems use three separate CRTs: a red, a green, and a blue tube. When the pictures from these three tubes are superimposed into a single picture, a color picture is produced. This is a Novabeam tube, based on the Schmidt optical system, developed by Henry Kloss.

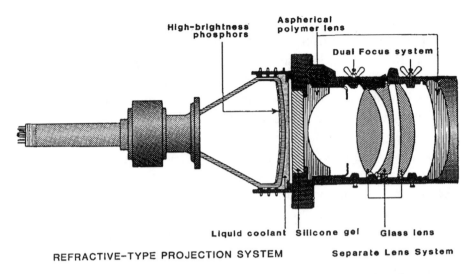

Figure 1-7. One type of separate-tube projection CRT is the refractive system.

based on the Schmidt optical system, developed by Henry Kloss. (*Figure 1-6.*) The second is the refractive system used by a number of other manufacturers. (*Figure 1-7.*) The Novabeam is the most efficient, but as with

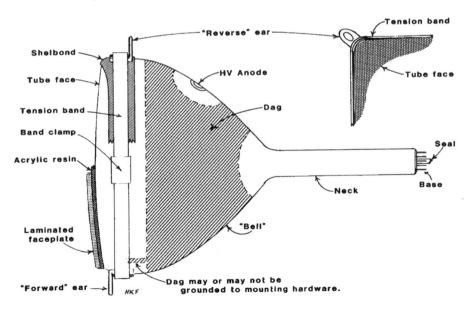

Figure 1-8. The nomenclature most commonly used when discussing a CRT.

all of the rest, has some disadvantages. It is presently used by only one manufacturer: Ampro. Several other manufacturers use the refraction system. Both systems are used in front and rear projection. Even though neither of them has a mask and each produces a solid red, green and blue picture, having no color dots, the picture has more pixels. It is still limited in clarity in the number of lines that make up the picture.

Projection TV sets have their place in large rooms for large audiences. They are more expensive and the CRTs have to be replaced more often. Except for the liquid crystal units, the tubes are designed to work at an exact distance from the screen. A curved screen produces a brighter picture, but narrows the viewing angle. Each system has its advantages and disadvantages. Available choices are a matter of personal preference and application.

Figure 1-8 lists the nomenclature most commonly used when discussing a CRT. This should help you tell your supplier just what you want without miscommunication.

Chapter Two

Understanding the CRT Numbering System

By the ES&T Staff
(Based on Sencore Tech Tip 145)

It's not absolutely essential to be familiar with component numbering systems, but it can help reduce errors in the ordering of a replacement part. This brief article explains the numbering systems for CRTs both before and since the introduction of the Worldwide Type Designation System (WTDS).

Most CRTs are registered according to some kind of industry standard. These standards define certain characteristics of the tube. Since April 1, 1982, these standards have been combined into a single worldwide standard. Before that date, several non-universal standards were used.

Old Standards

CRTs manufactured before April 1, 1982, were registered differently in the United States, Japan, and Europe. In general, the registration number broke down into three parts.

The first part of the old-type CRT number is a series of digits which signify the minimum diagonal viewing measurement of the CRT. For American tubes, this size is measured in inches. Thus a 19VACP22 would have a viewing area of 19 inches. Japanese tubes have this distance measured in millimeters.

The next part of the CRT number consists of one to four letters which designate a particular CRT within a group of CRTs having the same screen size. The final part of the CRT designation indicates the type of phosphor used. Black-and-white video CRTs use a "P4" designation for American listings or a "B4" designation for Japanese listings, while color CRTs use a "P22" (American) or "B22" (Japanese) listing. (*Figure 2-1.*) Computer CRTs or scope CRTs may use some other type of phosphor, and will have a different number following the "P" or "B." But, as you see, the "P" or "B" is not part of the tube designation.

Some CRTs have listings that do not have a "P" or "B" ending. The most common nonstandard ending is "TC01" or "TC02." These CRTs always have bonded yokes (or some other component) permanently attached to the CRT neck. The "TC__" ending simply indicates the type of yoke plug the CRT uses to connect to the chassis. The CRT is identical to one with a "P22" or "B22" ending. For example, a 15VAETC01 is identical to a 15VAEP22.

WTDS Standard

Since April 1, 1982, a new system for categorizing and numbering CRTs has been in use. This system is officially called Worldwide Type Designation System (WTDS) for picture tubes and monitor tubes. Until the adoption of this system, American, Japanese and European tube manufacturers had all numbered their tubes differently. This led to confusion and incomplete or inaccurate information. The WTDS number was an effort to simplify and unify CRT designations. (*Figure 2-2.*)

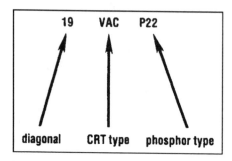

Figure 2-1. *Before April 1, 1982, the number used to identify picture tubes was different in the United States, Europe and Japan. Here is the number from an American CRT from that vintage. The first two digits identify the diagonal picture size, the second set of characters designates the CRT type, and the third set of characters designates the phosphor type.*

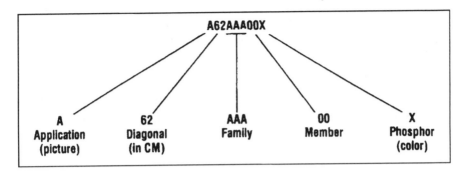

Figure 2-2. When the Worldwide Type Designation System (WTDS) was introduced in April of 1982, CRTs from all over the world were identified with the same six groups of symbols. One of the changes for U.S. manufacturers is that the diagonal size was now expressed in millimeters.

The WTDS number consists of six groups of symbols. The first symbol defines the application of the tube. This symbol is always a single letter; either an "A" for picture tubes or an "M" for video monitor tubes. A second group of symbols is a two-digit number that defines the minimum diagonal view. This measurement is always listed in centimeters. (1 inch = 2.54 centimeters.)

The next group of symbols consists of three letters that designate a family code for the CRT. Tubes within a particular family have specific mechanical and electrical characteristics. These letters are assigned alphabetically beginning with "AAA" followed by "AAB," "AAC," etc.

One or two digits follow the family code. These digits indicate a specific member within a particular family. A different member number would be assigned to tubes within the same family that have different neck diameters, for example. A single digit member symbol indicates a monochrome tube while a two-digit number indicates a color tube.

Following the one- or two-digit member symbol is the phosphor type designation. Color picture tubes are designated by the single letter "X," while color monitor tubes may have some other single letter designation. Monochrome picture tubes are designated by the two letters "WW." Other monochrome tubes, such as video monitors, have a different specific two-letter code to designate the phosphor type.

Some tubes contain integral neck components, such as bonded yokes. These tubes have a sixth group of symbols assigned to them, in which a two-digit number is used to define the characteristics of these integral neck components.

Chapter Three

Servicing Monochrome Televisions

By Dale Shackelford

Years ago it was not uncommon for electronics servicing technicians to service as many black-and-white televisions as they would color sets. Today, with the popularity of color sets, coupled with the extremely low cost of monochrome receivers, the monochrome set, like many other goods in this day and age, is considered disposable.

Technicians who still accept monochrome televisions for service realize that too much time spent on the diagnosis cannot be justified on the bottom line. The technician who can and will repair monochrome receivers in a matter of minutes, however, has additional sources of revenue that other technicians often let slip by.

Another reason to read an article on servicing of monochrome sets, even though they may frequently not be worth servicing, is that the diagnostic and servicing processes followed may be useful to know about when servicing more expensive sets.

Servicing the KTV Monochrome TV Set

Manufactured by Korean Electronics Co., Ltd., the 12-inch KTV monochrome receiver is fast becoming a popular consumer device. Two models of this set (the KT1210A and the KT1230) both use the exact same circuit boards, tuners and CRTs, though the KT1230 is capable of being powered by an external 12V DC source in addition to the standard 120V AC line voltage. Both models are equipped with internal line voltage transformers and standard (mono) earphone jacks.

Component Failure is Common

As is usually expected with most of the lower priced appliances, component failure in the two KTV models being discussed is quite common.
Fortunately, however, the same components seem to be causing the same
symptoms/problems in about 95 percent of the sets. When one works
with an average of five sets per week, all the same model and all exhibit-
ing the same symptoms, one becomes familiar with servicing them.

Noisy Volume Control is Chief Complaint

In most cases, the chief complaint with these sets is a scratchy or intermittent volume control. In an attempt to quickly rectify this problem, a
liberal amount of a brand name tuner cleaner was sprayed into R620 (potentiometer), making sure that the set was unplugged (the power switch
is connected to the pot), and that any run-off liquid was absorbed by a
shop rag. This particular brand of tuner cleaner, according to the claims
written on the side of the can, removes dirt, grease and oils from tuners,
potentiometers and other such components, and leaves a lubricating film
as well.

Because of these claims, and the success with this cleaner in other applications, I thought that a liberal spraying of the pot would resolve the
volume control problem and protect the component from future malfunctions. In the case of these sets, however, the tuner cleaner was not as
effective as I expected it would be.

After several experiments upon discarded variable resistors which are
typically used in these sets, I found that the pot must be sprayed out with
a strong solvent, typically used for the removal of flux, in lieu of completely dismantling the entire switch/potentiometer for a complete cleaning; a tedious, time-consuming process.

Although the practice of using strong solvents may not ordinarily be recommended for this application, a short spray of the pot with this solvent,

followed immediately by a liberal spraying with standard tuner cleaner, seems to prevent further problems. Apparently, flushing the solvent out of the pot immediately keeps it from causing excessive deterioration. You may want to experiment with other types of cleaners.

Once the potentiometer had been cleaned with the solvent/tuner cleaner combination, the customer had no more problems with scratchy or intermittent audio from the set. As a customer service, all such sets, regardless of the chief complaint, receive this treatment before the set is returned to the customer to prevent future problems.

Visible Scanning Lines

Another common complaint with the KT121OA and KT1230 models was described by the customer as "interference" lines that start at the bottom of the picture, moving slowly to the top. Once this line reaches the top of the screen, the picture rolls for a couple of seconds before the process repeats itself.

After observing the set for a couple of hours, the customer complaint was verified. What was described as interference, however, were actually scanning lines moving up the screen. After scoping the sync and vertical input circuits (Q301 and Q302 respectively) and finding the correct vertical sync pulses, I suspected that the problem was to be found in the low-voltage power supply.

In the case of most sets, the electrolytic capacitor used in the filtering of the AC line input would be a prime suspect, as it can have a tendency to dry up and require replacement. Since these models do not use the large filtering caps, I started with the bridge rectifier (*Figure 3-1*), the next stage of the low-voltage power supply.

I unsoldered one end of each diode in the bridge rectifier (D801 through D804) and checked to see if any were open or shorted. D804 was completely shorted, causing the same symptoms exhibited by a dried-up electrolytic capacitor.

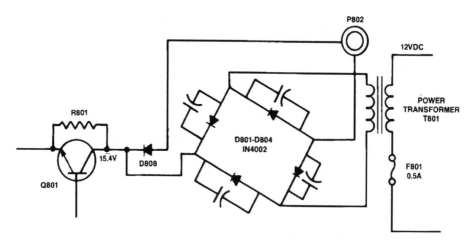

Figure 3-1. *Because the KTV does not have large filtering caps, the bridge rectifier was checked as the next stage of the low-voltage power supply.*

Replacing the entire rectifier bridge solved the video crawling problem. I replaced all the diodes rather than just the faulty one, because when one diode became faulty, additional stress was placed on the other diodes.

In sets which would later cross the bench with this video crawling problem, at least one of the diodes (though not necessarily D804) would be shorted, though D804 seemed to be the most often affected.

If the customer continues to operate the set with a bad diode in the rectifier bridge, a second and possibly third diode will go bad, causing F1 to open, resulting in a dead set by the time it arrives at the shop. Approximately 90 percent of the dead KT1210A and KT1230 models which come into the shop are due to an open line fuse (F1) as a result of bad rectifier diodes.

Retrace Lines

As with most other brands of television receivers, monochrome or color, KTVs are sometimes plagued by retrace lines over some or most of the screen, even though the video signal minus the retrace lines looks fairly good.

In most instances, retrace lines covering the entire screen would lead the servicing technician to check the video amp circuit. (See *Figure 3-2*.) In this case, the in-circuit voltage tests revealed 2.4V on the emitter of Q201 and approximately 2.8V on the base; pretty close to the desired voltages. At the collector of Q201, however, little or no voltage was present, where the schematic specified that there should be 72.2V. Voltage measurements were then made on the Q201 side of L202 (peaking coil), where the 72.2V should also be present, but it wasn't.

When I measured the voltage on the other side of L202, there was 72.2V present. Because the 72.2V was found on one side of the peaking coil, but not on the other (Q202) side, the inductor was removed from the circuit and tested with an ohmmeter. It was open. Replacing the peaking coil solved the recurrent full-screen retrace problems, and has solved the same problem in many other sets with the same symptoms since.

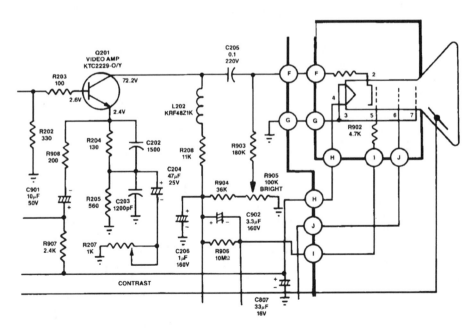

Figure 3-2. *In most instances, retrace lines covering the entire screen would lead the servicing technician to check the video amp circuit. In this case, the in-circuit voltage tests revealed 2.4V on the emitter of Q201 and approximately 2.8V on the base.*

Improper Vertical Deflection

The lack of full vertical deflection in any television set can be a nightmare for any technician, and these sets, though relatively simple, are no exception. In these KTV models, the vertical circuits are controlled by (and consist primarily of) a single IC package, identified in the schematic of *Figure 3-3* as Q302.

In cases where the top or bottom half of a raster is missing, the fault can usually be traced to the corresponding (top or bottom) vertical output transistor. When the set uses an IC package for both top and bottom vertical circuits, the simple replacement of the package will often solve the problem.

When this replacement does not affect the raster size, however, the diagnosis of the problem can become quite complex. This is especially true when the voltages on the IC correspond with those on the schematic, and attention must be turned to peripheral components such as biasing resistors and coupling capacitors.

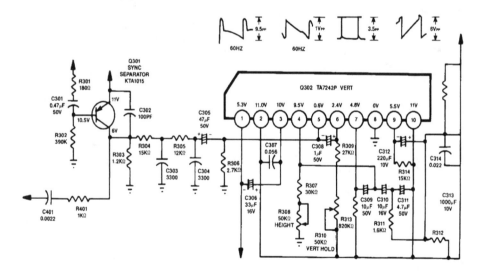

Figure 3-3. In KTV models, the vertical circuits are controlled by (and consist primarily of) a single IC package, identified in this schematic as Q302.

In one set that had only about four inches of raster in the center of the screen at the maximum height adjustment, the low-voltage power supply checked out good. Horizontal deflection was also fine. I carefully checked the voltages on Q302.

Minute variations in voltage on pins 5 and 6 were noted. Because C308 is tied between these two pins, I suspected that it might be the cause of the problem. I removed the capacitor from the circuit and checked it with a capacitance meter. The 1F, 50V capacitor measured approximately 0.2F; definitely outside tolerance limits. Replacing C308 restored the screen to full vertical deflection. From that time on, whenever I ran across one of these sets in which the picture was only a four-inch line across the center of the screen, I immediately replaced C308.

Experience Speeds Up Servicing

In every symptom described for this inexpensive model of monochrome television, the diagnosis was relatively straightforward. Although some shortcuts were taken in the repair of these sets, quite a bit of time was expended in diagnosing the individual problems, time which would have probably been more lucrative working on more expensive models, which, on the bottom line, would justify the amount of man-hours spent. On the other hand, once one case was diagnosed, subsequent cases were solved quickly.

Some technicians, as previously mentioned, simply refuse to accept small monochrome receivers for service, but when a valued customer (or a customer who may become loyal to your service center) walks in with that 12-inch portable KTV that they use in the basement or garage, you'll be ready for it.

Chapter Four

Technology Update: The Status of HDTV

By the ES&T Staff
(Based on Information from the Grand Alliance)

On May 24 of this year (1993), the three groups that had developed world-leading digital high-definition television (HDTV) (*Figure 4-1*) systems agreed to produce a single, best-of-the-best system to propose as the standard for the next generation of TV technology.

The three groups—AT&T and Zenith Electronics Corporation; General Instrument Corporation and the Massachusetts Institute of Technology; and a consortium of Philips Consumer Electronics, Thomson Consumer Electronics, and the David Sarnoff Research Center, are now all working together as the Digital HDTV "Grand Alliance."

The Grand Alliance creates a collaborative effort with a pool of technical talent and financial resources that should assure that North America is the first to deploy and benefit from this new digital technology.

In the past, the process of formulating an HDTV standard had concentrated on selecting the best system from among those proposed. Under the Grand Alliance, the best features of all the systems can now be combined to produce a system superior to that of any one of the individual proponents.

The Grand Alliance approach is expected to be good news for everyone—consumers, broadcasters, cable operators, and the computer and consumer electronics and telecommunications industries, as well as for North American workers. The proposal addresses the needs of these key constituencies and incorporates capabilities that are vital to each of them.

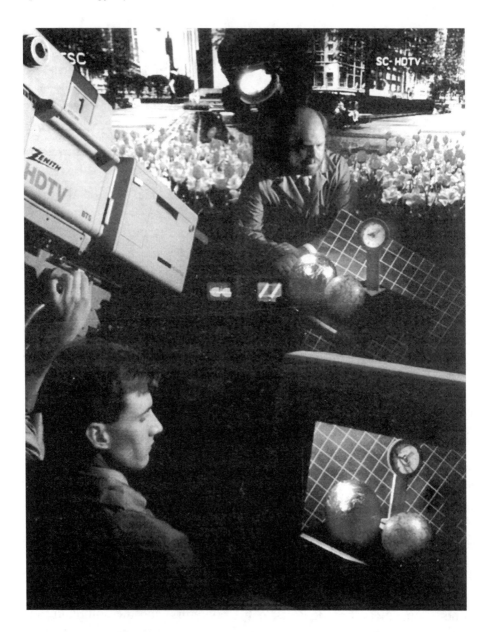

Figure 4-1. Researchers in one of Zenith's laboratories evaluate the high-definition picture performance of digital HDTV technology that is displayed here on a "flat tension mask" high-definition monitor.

For instance, the system incorporates progressive scan transmission capability and square pixel capability: two attributes that are important for promoting interoperability with computers and telecommunications. Likewise, concerns expressed by many broadcasters have been addressed by including interlaced scan transmission in the initial deployment.

The proposal will allow North America to maintain the worldwide technological lead it has established. The rapid adoption of an all-digital HDTV system in the United States, Canada, and the rest of North America, will promote the creation and maintenance of high-skilled jobs in the design and manufacture of HDTV receivers, displays, studio and transmission equipment, peripheral equipment, programming and software development, and semiconductor products. Consumers will reap the benefits of the best technical minds collaborating to bring noise-free, theater-quality pictures and sound to American homes, as well as a host of new applications in home entertainment, education, computer and medical imaging, factory automation and publishing—all stimulated by the early adoption of this technology.

The Process

The HDTV standard-setting process has been and will continue to be a public, open process. At the same time, it must proceed as rapidly as possible if U.S. and Canadian consumers and the countries' economies are to capitalize on this critical new technology. These are the next steps in the process:

(a) The Advisory Committee has reconvened its Technical Subgroup to evaluate the Grand Alliance proposal in detail. If necessary, this group may negotiate changes to the proposal system with the alliance members. In the meantime, the alliance members are finalizing the specifications of the combined system in a few areas that are not yet fully resolved.

(b) Once the Advisory Committee's Technical Subgroup has approved the basic concepts of the combined system, the Alliance members will work together to construct the system. After that, the Advisory Committee will conduct extensive laboratory tests in the U.S. and Canada to verify that the system meets its expectations. The Advisory Committee could then recommend the system to the FCC and simultaneously begin field test verification of the system's performance.

(c) The FCC, in turn, would consider the Committee's recommendation in a rulemaking proceeding which members of the alliance hope could be concluded by the end of 1994. Whatever standard is adopted, the FCC requires that the applicable technology be licensed to anyone on reasonable terms.

(d) It is anticipated that Canada and Mexico will simultaneously initiate similar, appropriate procedures to assure rapid adoption throughout North America. Moreover, because of early North American implementation, it is hoped that the rest of the world will adopt many of the elements of the North American HDTV standard.

Speed is of the essence. The Grand Alliance system, if ultimately accepted by the Advisory Committee and the FCC, will maintain and enhance the North American leadership position in digital television technology and in HDTV in particular.

Historical Perspective

The television we watch today uses the NTSC (National Television Systems Committee) standard, finalized in the late 1940s. While that standard has been improved, most notably by the incorporation of color in the 1950s, today's television is based on the same fundamental resolution parameters as the original service, including 525 horizontal lines and interlace scanning. The introduction of color television approximately 40 years ago was the last major advancement in the NTSC standard. The North American standardization activities were subsequently emulated throughout the world.

In the early 1980s, Japan's NHK proposed its MUSE HDTV interlaced system, based on 1,125 horizontal scan lines, and proposed its worldwide adoption. MUSE made the world aware of the goal of "high definition television," with quality equivalent to motion pictures, including a wide-screen format. The MUSE system renewed concerns in the United States about the capabilities of American technology. Many feared that American companies and employees would be shut out of a fundamental new technology.

In 1987, at the request of U.S. broadcasters, the FCC initiated its rulemaking on advanced television service and established its blue ribbon advisory committee for the purpose of recommending a broadcast standard. Former FCC Chairman, Richard E. Wiley, was appointed to chair this effort. Hundreds of companies and organizations from the U.S., Canada and Mexico have worked together within the numerous subcommittees, working parties, advisory groups and special panels of the Advisory Committee on Advanced Television Service (ACATS) during the past six years.

Several important steps followed:

(a) ACATS developed a competitive process by which proponents of systems were required to build prototype hardware which would then be thoroughly tested. This process sparked innovation and an entrepreneurial response: initially there were 23 proposals for systems submitted to ACATS in September 1988. (Hardware was actually built and tested for six systems.)

(b) The FCC made several key spectrum decisions that also helped spark innovation. The Commission decided in early 1990 that new ATV systems would share television bands with existing services and would utilize TV channels as presently defined. The Commission also decided that a simulcast approach, as first proposed by Zenith, would be followed. This meant that a new standard could provide a quantum leap forward from the current NTSC standard and would not be hindered by the requirements of the current standard, except to protect existing broadcast service during a period of transition.

(c) The FCC anticipated the need for interoperability of the standard with other media. Initially, the focus was on interoperability with cable television and satellite delivery; both were crucial to any broadcast standard. But the value of interoperability with computer and telecommunications applications became increasingly apparent, when the next technical advance came in the form of all-digital compression and transmission systems.

(d) Although the FCC had said in the Spring of 1990 that it would determine if all-digital technology was yet feasible, most observers viewed it as at least 10 years in the future. That same year, General Instrument became the first to announce an all-digital system, followed by the Philips-Thomson-Sarnoff consortium and by Zenith-AT&T. (Until then, there had been proposals for utilizing digital compression with analog transmission and proposals for hybrid digital/analog transmission.)

(e) Proponents later announced the use of packetized transmission, headers and descriptors, and composite-coded surround sound. (The Consortium and, to a lesser extent, Zenith-AT&T had previously adopted packetized transmission.) These features increase even further the interoperability of HDTV with computer and telecommunications systems. The introduction of all-digital systems had made such interoperability a reality.

(f) All-digital systems set the stage for another important step, which was taken in February 1992, when the Advanced Television Systems Committee (ATSC) recommended that the new standard include a flexible adaptive data allocation capability (and that the audio also be upgraded from stereo to surround sound).

Six systems (four of which were all-digital) underwent extensive testing in 1991 and 1992 at the Advanced Television Test Center (ATTC) in Alexandria, VA. Also participating in testing were CableLabs, which tested systems over a cable television test bed, and the Advanced Television Evaluation Laboratory (ATEL) in Ottawa, Canada. Canadian participation has been active and very important to the goal of creating a unified North American standard. The ACATS process has been critically dependent upon the unique, subjective picture-quality evaluations of the ATEL to ascertain if proponent systems truly result in "high-definition" pictures. Canadian participation has also been an invaluable part of the complex simulcast spectrum issues because of the thousands of miles of shared borders between Canada and the United States.

Following testing, the Advisory Committee decided to limit further consideration to those that had built the four all-digital systems: two systems proposed by GI and MIT, one proposed by Zenith and AT&T, and one proposed by Sarnoff, Philips and Thomson. The Advisory Committee decided that while all of the digital systems provided impressive results, no winner could then be proposed to the FCC as the U.S. standard. The Committee ordered a round of supplementary tests to evaluate improvements that had been made to the individual systems.

Speeding HDTV Implementation

The formation of the Grand Alliance has eliminated the need for another round of testing on the individual systems, the results of which could have been inconclusive. Thus, the formation of the Grand Alliance could save a year or more in the implementation of HDTV by reducing the risk of inconclusive test results and the possibility of legal or other challenges.

If accepted by ACATS and by the FCC, the system will speed the implementation of HDTV in the United States. That, in turn, should set the stage for adoption in Canada, enabling North America to maintain and enhance its worldwide lead in the development of this vital technology.

Interoperability

Representatives of the computer industry have made significant contributions to the HDTV standards process and to the Grand Alliance systems. They participated in the work of ACATS and helped to articulate the need for features that could enhance the interoperability of an all-digital system. The standard will be better than it would have been thanks to their participation. It is important to recognize the extent of the commitment being made to increase interoperability of HDTV with computers and television. Participants from non-broadcast industries suggested a number of significant features for the standard which have been incorporated into the Grand Alliance proposal:

(a) They sought an all-digital advanced television standard. The proposal is for an all-digital system.
(b) They said that the digital data stream should have a prioritized and packetized data support structure. The Alliance proposal incorporates such a structure.
(c) They maintained that the standard should include source adaptive coding. The Alliance proposal does.
(d) They requested that the standard provide for square pixels to facilitate computer graphics. The Alliance proposal provides for square pixels.
(e) They requested that the standard utilize a progressive scanning format. The Grand Alliance proposal includes progressive scanning from the outset and promises a migration plan to the eventual exclusive use of progressive scanning.

Other aspects of the Grand Alliance system enhance interoperability with computers and telecommunications. The Grand Alliance system is very similar to the evolving MPEG-2 compression approach, which is currently in working draft status in the MPEG Committee of the International Standards Organization.

The Alliance compression system may also include additional capabilities in order to assure the highest picture quality possible. If so, the Grand Alliance will endeavor to get these capabilities incorporated in the MPEG standard.

Another aspect of the Grand Alliance system which enhances interoperability is the fixed-length packet format that provides for flexible delivery of video, audio, text, graphics and other data by broadcast, cable, satellite and fiber. This packet data format provides flexibility and a high degree of interoperability with other emerging telecommunications and data networks that use similar technology, such as Asynchronous Transfer Mode (ATM), the emerging standard for broadband telecommunications networks. Finally, the proposal's packetized data transport structure utilizes universal headers and descriptors in order to provide flexibility and extensibility (i.e. headroom) for future growth of system capabilities.

More to Come

The introduction of a completely new television broadcast system will mean a whole new technology for consumer electronics servicing technicians to learn. As the HDTV system evolves toward fruition, ES&T will continue to provide technical detail on the circuitry

Technology:
The Digital Micromirror Device
By the ES&T Staff
(Based on Information from Texas Instruments)

While the Grand Alliance continues its development work on the HDTV system, other organizations continue to develop the technology that may one day be used to display an HDTV picture. One such display unit is the digital micromirror device (DMD) (*Figure 4-2*), currently being developed at Texas Instruments.

The DMD

The DMD is a micromechanical, reflective spatial light modulator monolithically fabricated over a conventional CMOS SRAM address circuit. A 768 x 576 pixel array DMD (442,368 mirrors) has been developed for and demonstrated in one-DMD and three-DMD projection television systems that are capable of projecting pictures with diagonals ranging in size from 42 inches to 13 feet.

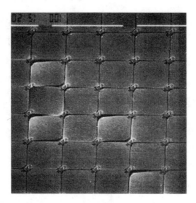

Figure 4-2. The deformable mirror device (DMD), developed by Texas Instruments, may one day compete with the LCD and the CRT as display elements for HDTV. In this early version of the DMD, the contrast ratio was degraded because of stray light diffracted from the hinges, hinge support posts, and the mirror edges at the landing tip which are not at right angles to the mirror surface. This stray light adds to the light that is projected by mirrors that are in the off state.

The DMD as a Potential Video Display Device

Most consumer projection television (PTV) sets sold today are CRT rear-projection systems. The display portion of these products consists of three CRTs, one for each of the primary colors. If other technologies such as the liquid crystal display (LCD) and the digital micromirror device (DMD) are to compete, they will have to have performance and reliability that meet the standards of the CRT, or be superior, and will have to cost about the same in order to be accepted in the PTV marketplace.

CRTs are the workhorse of display technology, and are constantly improved to meet higher performance standards of resolution, brightness, contrast ratio, and convergence in PTV applications. However, CRT performance is expected to soon reach its limit.

With the anticipated advent of digital, high-definition PTV in the late 1990s, with twice the resolution of standard-definition PTV, CRTs may not be able to meet the demanding standard of consumers for image quality that is comparable, or nearly so, to what they are used to seeing at the movie theaters.

Developers are therefore making large investments in LCD technology for PTV. LCD technology is currently faced with the challenges of improving its optical efficiency in order to achieve increased brightness, and with the development of high-yield active-matrix address circuits based on amorphous silicon or polysilicon technology.

According to Texas Instruments, the DMD provides a promising alternative to LCD technology in PTV applications, with performance equal to or better than that of a CRT, and with the potential for being cost competitive with the CRT. Currently, the DMD has more optical efficiency than conventional LCDs.

The DMD Chip

The DMD was invented in 1987. It is the outgrowth of work that began a decade earlier at Texas Instruments on micromechanical, analog light modulators. These were called the deformable mirror device (also DMD). The DMD is a reflective spatial light modulator, consisting of an array of tiny aluminum mirrors that can rotate. The mirrors are monolithically fabricated over an address circuit consisting of conventional CMOS SRAM cells. The mirrors are 16m wide on a pitch of 17m, and are capable of rotating $\pm$10 degrees.

The mirror is suspended over an air gap by two thin torsion hinges supported by posts that are electrically connected to an underlying bias/reset bus. This bus interconnects all the mirrors directly to a bond pad so that a bias/reset voltage waveform can be applied to the mirrors by a circuit that is not on the same chip. Underlying the mirror are a pair of address electrodes that are connected to the complementary side of an underlying SRAM cell.

Depending on the state of the SRAM cell, the mirror is electrostatically attracted by a combination of bias and address voltage to one or the other of the address electrodes. It rotates until its tip touches a landing electrode held at the same potential as the mirror.

A "1" in the memory cell causes the mirror to rotate +10 degrees. A "0" in the memory causes the mirror to rotate -10 degrees. Although the DMD can be operated in an analog mode, the mirrors are biased in such a way that only the digital landing states of ± 10 degrees are possible. The digital mode of operation permits the use of low-voltage CMOS and ensures large, uniform deflection angles.

A 768 x 576 DMD pixel array (442,368 mirrors) has been developed for PTV applications. Such an array meets the European resolution standard, and exceeds the current United States resolution standard.

The DMD superstructure is formed on top of the CMOS, using four photolithography layers. The three metallization layers—the electrode, the hinge, and the mirror—are sputter-deposited aluminum structures that are plasma (dry) etched. During the fabrication process, a spacer made of an organic material is formed between the SRAM and the mirror structure. Later the spacer is destroyed, creating an air gap between the address electrodes and mirror.

The Projection System

The ± 10 degree mirror rotation angles are converted into high-contrast brightness variations by the use of a darkfield projection system. In a darkfield projection system, the light beam is aimed at an angle to the object whose image is to be projected. Only portions of the object that are at a certain angle to the incident light reflects light through the projection lens to form an image on the screen. All other light from the light source is reflected away from the projection lens.

DMD projection systems based on a single chip, and DMD projection systems based on three chips, have been demonstrated. In the three-chip system, one chip is used for each of the primary colors. (R, G, B.) This is similar to most LCD projection displays. Because of the DMD's efficient use of light, it is also practical to use a single-chip system. The single-chip system is less costly and does not require convergence.

In the single-chip scheme, a single DMD is illuminated with the primary colors in a timed sequence using a color wheel. (*Figure 4-3.*) In either system, light from a metal-halide or xenon arc lamp, or similar source, is aimed through a condenser lens and directed at an angle of +20 degrees from the normal of the DMD and orthogonal to the rotation axes of the mirrors. A projection lens positioned above the chip (or chips) produces an enlarged image of each DMD mirror on a projection screen.

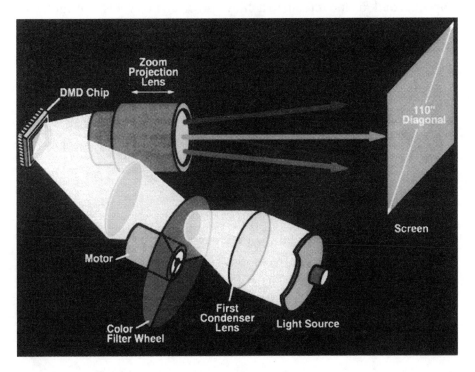

Figure 4-3. *The DMD is used in a darkfield projection system. Projection of color TV using the DMD would be accomplished either by using three DMDs and three-color projectors, or by using a single DMD that would be time sequentially illuminated with different colored lights using a color wheel system.*

If one of the mirrors is rotated to +10 degrees (memory address "1") it reflects the incoming light into the pupil of the projection lens and the mirror appears bright (on) at the projection screen. If a mirror is rotated to -10 degrees (memory address "0") it reflects light at an angle of -40 degrees with respect to the projection lens, so that the light from it does not arrive at the screen. This mirror appears dark (off) at the projection screen. The flat structures (post tops, hinges) reflect light at an angle of -20 degrees. This light also misses the pupil of the projection lens and is not projected onto the screen.

The mirrors can change position very quickly: in about 10 sec. The DMD can take advantage of this capability and vary the amount of time during which the mirror reflects light onto the screen versus the amount of time it appears dark on the screen, to achieve precise gray levels of brightness. This grayscale variation is accomplished by the pulsewidth modulation of the mirrors.

Each video field is subdivided into time intervals, or bit times. Each interval is one-half as long as the preceding interval. During each of these bit times, the mirrors are addressed by the underlying SRAM array to be in either the on or off state. If an 8-bit modulation scheme is chosen, any one of 28 or 256 gray levels is possible.

Improved Performance DMD

To be competitive with traditional CRT-based PTVs, the DMD projection system must have a contrast ratio (CR) of better than 100:1, and still maintain optical efficiency. The conventional DMD superstructure can only deliver a CR in the neighborhood of 50:1 with an f/2.8 projection lens. An f/2.8 lens is the fastest lens that preserves the darkfield imaging condition for mirror rotation angles of ±10 degrees.

Degradation in the contrast ratio of the conventional DMD superstructure is for the most part caused by stray light diffracted from the hinges, hinge support posts, and the mirror edges at the landing tip which are not

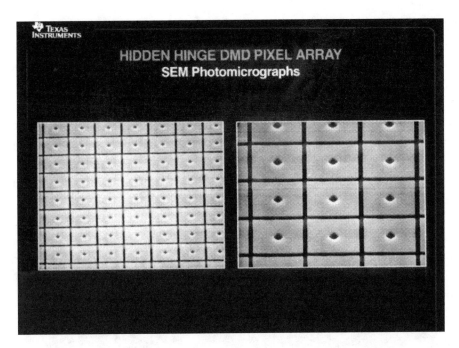

Figure 4-4. A new DMD superstructure is called the hidden hinge DMD. Viewed from the top, the pixel array looks like a set of closely packed square mirrors. This structure offers nearly ideal optical characteristics in terms of both contrast ratio and optical efficiency. The hinges and hinge supports in this system are hidden under the mirror.

at right angles to the mirror surface. This stray light adds to the light that is projected by mirrors that are in the off state. Together with other sources of stray light, including lens glare, the CR degrades to unacceptable levels.

A new DMD superstructure, *Figure 4-4*, is called the hidden hinge DMD. Viewed from the top, the pixel array looks like a set of closely packed square mirrors. This structure offers nearly ideal optical characteristics in terms of both contrast ratio and optical efficiency. The hinges and hinge supports in this system are hidden under the mirror. (*Figure 4-5*). The mirror is connected by a mirror support post to an underlying yoke. The yoke is in turn connected by torsion hinges to hinge support posts. The address electrodes and landing electrodes are coplanar with the hinge.

There are two air gaps, one between the mirror and the underlying hinges and address electrodes, and a second between the coplanar address electrodes and hinge, and an underlying third level of metal of the CMOS SRAM structure.

The hinges and support posts are hidden under the mirror, and therefore cannot diffract light, and so do not degrade the CR. Because the mirror edges at the landing tip are perpendicular to the mirror surface, the landing tip diffracts less light, leading to less degradation in CR. The greater mirror area leads to greater optical efficiency.

A 768 x 576 hidden hinge DMD has been tested in an f/2.8 projector using a test pattern to properly take into account all sources of system CR degradation. System contrast ratios of 110:1 have been measured, meeting the contrast ratio requirement of better than 100:1 for PTVs

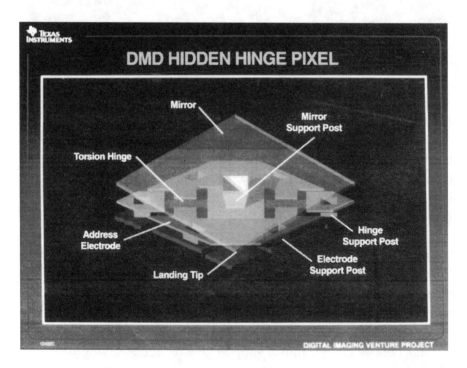

Figure 4-5. This view shows the detail of a single pixel in the DMD hidden hinge system.

with no sacrifice in optical efficiency. In fact, the efficiency increases because of the higher percentage of active mirror area.

Will This be the Display of the Future?

Service technicians are most familiar with CRTs as video displays. More recently, they have had to become familiar with LCDs. The deformable mirror device is an entirely new display system that could one day displace both CRTs and LCDs.

Chapter Five

Troubleshooting the Philips S1/S4 Chassis

By Dale Shackelford

It is not uncommon for a servicing technician to work on a TV set one week, and the next week see the exact same chassis, cabinet style and circuitry in another set, with a different brand name stenciled across the front. Sometimes several schematics have been purchased for a series of products that are exactly the same, yet have different model/manufacturer identification.

The S1 and S4 chassis manufactured by the Philips Consumer Electronics Company are classic examples of the same chassis used in a wide variety of models with different brand names. These chassis are used in no fewer than six different models, with screen sizes ranging from 13 inches to 27 inches. (*Table 5-1.*)

Every one of these sets may be serviced using Technical Manual No. 7502, available from Philips. This comprehensive manual has 72 pages of valuable schematics and information on various aspects of the S1/S4 chassis, including designs and options such as AVIO (audio/video input/output) jacks, picture in picture (PIP) modules and stereo sound. *Figure 5-1* is the block diagram of this chassis, *Figure 5-2* is the wiring diagram, and *Figure 5-3* illustrates the component layout.

The Power Supply

The power supply of the S1/S4 chassis is relatively straightforward, with a full-wave (diode) bridge rectifier, filtering capacitors, and a voltage regulator IC (IC430) that injects 155V DC into the system.

S1 MODEL TO CHASSIS LIST

Model	Screen Size	Chassis Reference	Model	Screen Size	Chassis Reference
Magnavox			**Magnavox (continued)**		
HD1920	19"	19S101	CP4581	25"	25S107
CMX162 (HOT.-MOT.)	19"	19S101	XML192 (CANADA HOT.-MOT.)	25"	25S108
EMX162 (HOT.-MOT.)	19"	19S101	XR2542 (CANADA)	25"	25S109
XMX162 (CANADA HOT.-MOT.)	19"	19S101	XS2560 (CANADA)	25"	25S110
RR1930	19"	19S102	XS2555 (CANADA)	25"	25S111
XR1930 (CANADA)	19"	19S102			
RR1936	19"	19S102	RS2655	26"	26S101
RR1937	19"	19S102	HD2650	26"	26S102
RR1938	19"	19S102	RS2655	26"	26S103
RR1940	19"	19S104	HD2650	26"	26S104
XR1940 (CANADA)	19"	19S104			
RR1944	19"	19S104	HD2762	27"	27S101
RS1960	19"	19S105	RK5520	27"	27S101
HD1925	19"	19S107	RS2745	27"	27S101
XD1925 (CANADA)	19"	19S107	RS2755	27"	27S101
RS1965	19"	19S108	RK5521	27"	27S101
RS1970	19"	19S109	RR2740	27"	27S102
RP1945	19"	19S112	XD2762 (CANADA)	27"	27S104
EM2019 (HOT.-MOT.)	19"	19S115	XS2745 (CANADA)	27"	27S104
GM2019 (HOT.-MOT.)	19"	19S115			
			Sylvania		
CMK172 (HOT.-MOT.)	20"	20S101	SRC194	19"	19S102
EMK172 (HOT.-MOT.)	20"	20S101	SRW195	19"	19S104
XMK172 (CANADA HOT.-MOT.)	20"	20S101	SSB195	19"	19S109
RR2040	20"	20S103			
RS2045	20"	20S104	SRW202	20"	20S103
HD2038	20"	20S105	SSB206	20"	20S105
RS2002	20"	20S105			
RS2050	20"	20S105	SSW256	25"	25S101
RS2080	20"	20S105	SRA453	25"	25S102
XS2050 (CANADA)	20"	20S105	SRP453	25"	25S102
XS2080 (CANADA)	20"	20S105	SRW252	25"	25S102
XD2038	20"	20S105	SSA454	25"	25S104
			SSP454	25"	25S104
RS2560	25"	25S101	SPP456-22XX	25"	25S107
RS2563	25"	25S101	SPA456-12XX	25"	25S107
RS2566	25"	25S101	SPC258	25"	25S107
XD2503 (CANADA)	25"	25S101	SSC263	26"	26S101
XS2563 (CANADA)	25"	25S101			
XS2566 (CANADA)	25"	25S101	SRW273	27"	27S102
HD2504	25"	25S101			
CR4510	25"	25S102	**Crosley**		
CR4520	25"	25S102	CT1911	19"	19S102
CR4521	25"	25S102			
CX9512	25"	25S102	CT2031	20"	20S103
CX9514	25"	25S102	CT2051	20"	20S105
CX9516	25"	25S102			
HD2501	25"	25S102	CT2511	25"	25S102
RR2540	25"	25S102	CC2533	25"	25S102
HD2502	25"	25S102	CC2534	25"	25S102
RR2544	25"	25S102	CC2535	25"	25S102
CR4522	25"	25S102	CC2557	25"	25S104
RR2535	25"	25S102	CC2558	25"	25S104
RP2545	25"	25S103	CC2545	25"	25S104
CS4535	25"	25S104	CT2521	25"	25S110
CS4536	25"	25S104			
CX9522	25"	25S104	CT2725	27"	27S102
CX9526	25"	25S104			
RS2555	25"	25S105	**Philco**		
CML192 (HOT.-MOT.)	25"	25S106	P2010R	20"	20S103
EML192 (HOT.-MOT.)	25"	25S106			
RP2575	25"	25S107	P2510R	25"	25S102
RP2576-C1XX	25"	25S107	P2505R	25"	25S102
CP4580	25"	25S107			
RP2577	25"	25S107	P2702R	27"	27S102

Table 5-1. *The S1 and S4 chassis are used in no fewer than six different models, with screen sizes ranging from 13 inches to 27 inches*

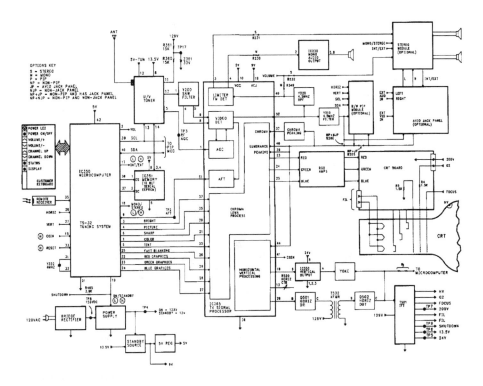

Figure 5-1. *The block diagram of the S1/S4 chassis.*

As with most modern television receivers, these chassis incorporate a "standby" circuit which is activated when the set is turned off. This circuit has an output of approximately 9V DC, which is used to power the IR receiver circuitry for remote control power up, microcomputer reset and PIP control. Thus, whenever the set is plugged into a 120V AC source, the set is either in an "on" or "standby" condition.

One of the most important components for the servicing technician to be aware of when servicing these chassis is R407. (*Figure 5-4.*) This 100W, 1/3W resistor is fusible, and when it fails it is due to a problem within the horizontal section.

When the set is turned on, 13.5V DC is fed from the flyback transformer (T501) to the anode of D045, which ultimately reverse biases Q403. If D405 is open, or a problem develops within the horizontal section that

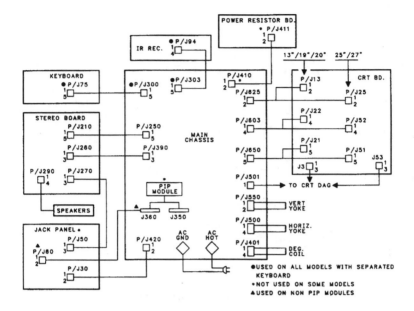

Figure 5-2. *The wiring diagram of the S1/S4 chassis.*

results in the 13.5V DC being absent from the emitter of Q403, R407 will be destroyed when the 129V DC source comes up while Q403 is still in the on condition. (*Figure 5-4.*)

If you encounter a destroyed R407 in one of these chassis, connect an external 13.5V DC source to the emitter of Q403 and ground the base of Q402 (power on/off transistor). Then connect the AC power cord to a variable transformer set to 0V, and a scope probe to pin 6 of T501.

Slowly increase the voltage output of the variable transformer. You should observe a horizontal pulse from pin 6 once you reach approximately 50V AC. If this pulse is not present, troubleshoot the horizontal while the set is powered from the reduced-voltage AC source. Once the problem is remedied, the pulse will be present.

On the opposite side of the coin, too high a voltage on the base of Q402 will keep the transistor from turning off, and the set will not operate. This excessive voltage is usually generated by a typical shutdown circuit.

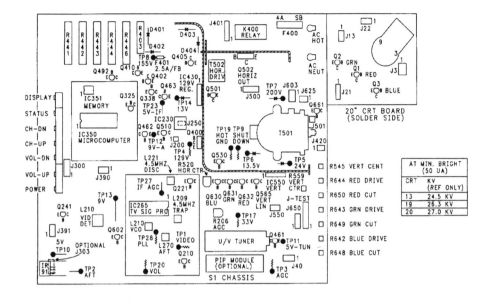

Figure 5-3. *The component layout diagram of the S1/S4 chassis.*

Microcomputer Control

With the computerization of recent television receivers, it is not uncommon to find entire logic systems contained entirely within two or three integrated circuits. The S1/S4 chassis are no exception. For example, the tuning system is controlled by a microcomputer and a memory IC (IC350 and IC351, respectively) in addition to the U/V tuner. With this combination, this system is capable of receiving up to 178 channels, while providing an on-screen display for all consumer adjustable controls such as brightness, color, volume and sleep timer.

When problems develop in the tuning system or any of the on-screen displays, check the voltages both into and out of the microcomputer (IC350), the memory IC (IC351) and the U/V tuner. (*Figure 5-5.*)

The voltages on the U/V tuner can be informative, as this unit is powered from three separate supplies. The 5V DC (tuner) voltage (pin 12) is de-

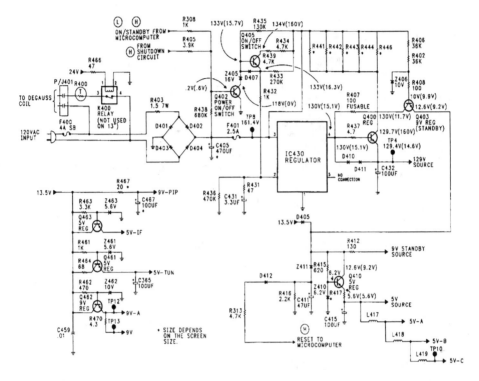

Figure 5-4. One of the most important components for the servicing technician to be aware of when servicing these chassis is R407.

rived from the emitter of Q461, a voltage regulation transistor which derives its voltage from the 13.5V DC, and as with the voltage at pin 12, the voltage is scan derived.

Unlike pins 6 and 12, however, the 33V DC tuning voltage is taken from the 129V DC source via R360, R361, and R366. If any "on" of these resistors opens, or the 33V zener diode (Z361) shorts, the 33V DC needed to power the tuner would obviously not be present.

In the event that all voltages are correct, check the main tuning system oscillator to determine if the requisite 4 MHz signal is present on pins 31 and 32 of IC350. Additionally, pins 28 and 29 are connected to the 6.5 MHz tank circuit/oscillator necessary for on-screen display of various functions.

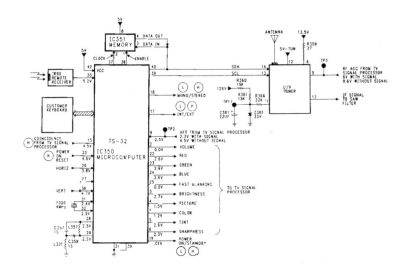

Figure 5-5. *When problems develop in the tuning system or any of the on-screen displays, check the voltages both into and out of the microcomputer (IC350), the memory IC (IC351) and the U/V tuner.*

If the tuning system and adjacent circuitry, including voltages and frequencies check out, yet the system does not operate correctly, the microcomputer (IC350) may have to be replaced. Because of the various designs and options, this IC will be different in different sets. Look closely at the original chip. If the chip has TOSHIBA printed on it, and the particular set does not employ the PIP option, the part number of the IC is 4835-209-17352.

If the receiver does feature the PIP option, and IC350 is marked TOSHIBA, the part number of the IC is 4835-209-47111. If the original IC350 is printed with the name SIGNETICS, contact Philips' technical department for assistance:

Philips Service Company, Parts Order Department
112 Polk Street, PO Box 967
Greenville, TN 37744
Phone: 800-851-8885 or FAX: 800-535-3715.

If a loss of stored data occurs, it is likely that the memory chip (IC351) will have to be replaced. If IC 351 has TOSHIBA printed on it, the part number is 4835-209-17315. If the name SIGNETICS appears, the part number is 4835-209-47106. In either case, take precautions when handling these ICs, and protect them from electrostatic discharge damage.

Using the Test Mode

One of the advanced features of this microcomputer/memory system is the test mode. This test mode allows servicing technicians to make various adjustments, via the remote control, to the brightness, picture, color tint and sharpness parameters. Thus, you can set up the set without even opening up the cabinet.

To enter the test mode, use an appropriate remote control unit that features a numerical input pad, and enter the following sequence: 0-6-2-5-9-6-MENU. This sequence must be entered fast enough that the on-screen display of the numbers does not time out between entries, although the display will register only one or two of the digit entries at a time. Once you have successfully done this, you should see a display at the bottom of the screen that looks like this:

TS32-C2   B300
5 C   11   1F

The TS32 on the display denotes the identification of the tuning system which is currently used in all receivers using the S1/S4 chassis configuration; from the 13-inch to the 27-inch models. The TS32 system, in addition to providing an excellent quality picture, is also capable of controlling PIP functions and stereo sound, if these options are present within the set.

The C immediately following the TS32 indicates that the particular receiver is designed for consumer use. If the M or H were to be displayed (rather than the C) it would indicate that the receiver contained a special

Motel or Hospital package. The 2 following the consumer designation indicates the version of the software used in this particular set.

If a letter appears immediately after the software version, it indicates that there is an error in one of the subsystems. For example, if the display were to read TS32-C2M, an error in the memory subsystem is indicated. If the letter T or P were displayed, the error would be found in the tuner or PIP subsystems respectively.

The B300 shown to the far right is the internal timer, showing minutes in hexadecimal form. This number will most likely be a different value when you view it.

This timer is used primarily for the sleep timer function, which allows the consumer to program the set to turn off automatically after up to 120 minutes, in 30-minute increments. The on-screen display even wishes the viewer a "Good Night" during the 15 seconds of operation, during which time a second-by-second countdown is displayed.

On the next line, at the extreme left, is the channel to which the receiver is currently tuned. The C on the second line indicates that the set is currently in the SERVICE mode. The letter A or B would indicate that the receiver is in a factory set-up mode.

In the SERVICE mode, unlike either of the factory modes, the registers (adjustments) may be set in small increments, from minimum to maximum. In the factory modes, the registers can only be set at minimum, medium or maximum, with no incremental values possible. To go from the service setting to either of the factory modes (A/B), press menu or the up/down arrows on the remote control unit.

When the display is changed to Factory (A), the receiver will default to channel 2. Factory setting (B) will default the receiver to channel 3, and Service setting (C) will default the receiver to the channel to which it was tuned prior to entering the test mode. In all settings, the channels may be changed via the remote control unit by using the up-down arrow function keys, as the numerical keys will only change the register values.

The next number in the series, in this instance 11, indicates the register is currently engaged. The set of numbers, or number/letter combination to the far right denotes the current value of the register displayed at the immediate left, in hexadecimal notation.

Thus, if the register number is 15, sharpness, and the value is 1F, the value of the sharpness register is set to factory specifications or mid-level. If the value of the register were to indicate 0, the sharpness register would be at a minimum.

To select a particular register, press 9 and the last number of the register desired, or simply press Reset (or the up-down arrows) on the remote until the desired register appears.

To change the value of the register, press + or - (left-right arrows) in the Service mode (C), as previously mentioned, the register values may be changed in small increments. In Factory (A or B) modes, the same keys will move the register values from MIN-MID-MAX only.

To save any changes made in any of the register values, the television receiver must be turned off at the set. If the receiver is turned off with the remote control, no changes will be saved.

This Set Has a Hot Chassis

As with all appliances that utilize a hot chassis, an isolation transformer must be used when servicing these chassis.

With a little time invested in the understanding of the S1/S4 family of chassis, you will become familiar with several different models and brands which you may never have heard of, much less repaired. While this chapter is not comprehensive enough to cover any problem which may be encountered with sets containing these chassis, it will give you a place to start. The rest, as they say, is up to you.

Chapter Six

Understanding TV Horizontal Output/Deflection Circuits

By Glen Kropuenske

To many technicians, the operation of the horizontal output stage and flyback transformer of a TV set is mysterious. Few technical books provide a clear explanation of how the stage operates. Therefore, when the stage develops problems, as it often does, technicians have difficulty interpreting the waveform in this circuit and relating symptoms to possible causes.

This article describes the operation of a horizontal output circuit and examines the currents and voltages within the stage. Later we'll examine what measurements you can make in the horizontal output stage and how to use those measurements to isolate defects.

Key Horizontal Output Stage Components

The horizontal output circuit consists of six key components:

(a) Horizontal output transistor (Q1).
(b) Flyback transformer (flyback).
(c) Retrace timing capacitor or "safety cap" (Ct).
(d) Damper diode (D1).
(e) Horizontal yoke.
(f) Yoke series capacitor (Cs).

Figure 6-1 shows a simplified horizontal output stage with these six components.

Although there are many different TV and monitor chassis, the horizontal output stages operate virtually the same in all of them. The stage is ener-

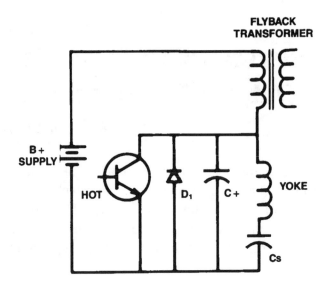

Figure 6-1. The horizontal output circuit consists of six key components.

gized by current from a B+ power supply flowing through the primary winding of the flyback transformer. The path for current is provided by the conduction of the horizontal output transistor (HOT).

Once energized, the horizontal output stage switches between two resonant LC circuit conditions. The resonant LC circuits are formed by the flyback, yoke and capacitors in the horizontal output stage. Currents alternating in the stage produce sawtooth currents in the yoke and primary winding of the flyback transformer. The sawtooth current in the yoke provides the magnetic field that deflects the electron beam across the CRT screen, thus painting the TV picture. Let's look at the role that each component plays in the operation of the horizontal output stage, then put them together and analyze the circuit.

Horizontal Output Transistor (HOT)

The horizontal output transistor, or HOT, serves as a switch. It switches on and off at the horizontal scan frequency of the TV or monitor. When switched on, it provides a low resistance path for the flyback primary and yoke currents. When switched off it serves as an open circuit.

Like a class C amplifier, the horizontal output transistor has no DC bias applied to its base. The transistor is switched on by the drive signal applied to the base/emitter junction. Because the horizontal output transistor is a power transistor, it is base current (resulting from the drive) that actually controls the transistor switch.

One of the biggest misconceptions about the horizontal output stage is that, when the horizontal output transistor is on, many mistakenly believe that the horizontal output transistor is conducting when the flyback pulse is produced at the collector (retrace or sync time). This is not the case. The horizontal oscillator and driver stage are synchronized to turn the horizontal output transistor on prior to horizontal sync or retrace. In a TV receiver, the transistor turns on 30S to 35S before horizontal sync or retrace. The transistor is switched off at the start of horizontal sync. (See *Figure 6-2.*)

Because a power transistor is not an ideal switch, several factors can influence the operation of the horizontal output stage. First, for proper operation, the base drive current must bias the transistor into a com-

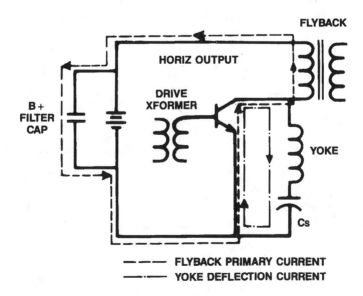

Figure 6-2. Horizontal output transistor conduction current paths.

pletely "on" state, meaning low collector-to-emitter resistance. Insufficient drive leaves resistance between the emitter and the collector, which resists current flow and generates high transistor heat.

Equally critical is the time it takes to switch the horizontal output transistor between on and off states. As the transistor switches, the emitter to collector resistance changes from less than 5Ω (on) to greater than 10MΩ (off). Current flowing through the transistor during the transitions produce heat. Longer transitions result in higher transistor heating. Switching irregularities can cause the horizontal output transistor to develop excessive heat and fail.

Second, transistor theory dictates that the amount of collector current that will flow through the transistor is determined by multiplying the base current times the current gain (beta) of the transistor. Unlike other switching transistors, the base current may be considerable (100mA to 300mA) to deliver the needed collector current peaks. The transistor must have proper gain, and the horizontal driver stage and driver transformer must produce adequate base current. Reduced beta or insufficient base drive current limits the collector current, starving the flyback and yoke of the current needed to produce full deflection and high voltage. (*Figure 6-3.*)

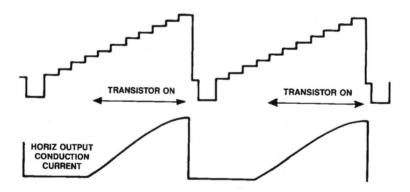

Figure 6-3. Conduction time of the horizontal output transistor relative to one horizontal line of video.

Flyback Transformer

The horizontal output transformer is called the flyback or integrated high voltage transformer (IHVT). An IHVT is a flyback transformer that includes the high voltage multiplier. The flyback is primarily responsible for developing high voltage. It is constructed with a powdered-iron or ceramic core to work efficiently at high frequencies.

The flyback transformer includes one primary winding and many secondary windings. (*Figure 6-4.*) The main secondary winding supplies voltage pulses to a voltage multiplier. Other secondary windings supply CRT filament power, keying pulses and scan-derived power supplies.

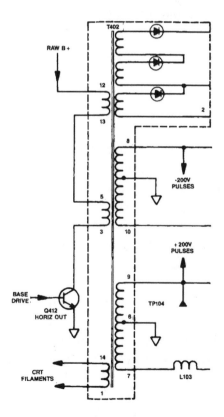

Figure 6-4. *The flyback transformer includes one primary winding and many secondary windings.*

The horizontal output stage develops high voltage pulses to the individual windings of the flyback transformer. This is accomplished as a result of the inductive nature of the flyback transformer primary, along with timing capacitors and the horizontal output transistor. To understand how these pulses are produced you need to recall some basic inductor theory.

Inductor theory tells us that the voltage induced across any inductance is always proportional to the rate of change of current with time. This means that a fast current change in an inductor can produce a large induced voltage. This large induced voltage is known as an inductive "kick."

The mathematical formula that gives the relationship between the voltage across and inductor, and the current through it, is:

$$VL = L(di/dt)$$

where L is the inductance and di/dt is the expression for the rate of change of the current through the conductor at any instant.

During the conduction time of the horizontal output transistor, there is an inductive or linear rise in current in the flyback primary. This increase in current produces a constant induced voltage across the flyback windings. When the horizontal output transistor is abruptly turned off current flow in the flyback primary abruptly ends. The magnetic field within the flyback core collapses rapidly, producing a high induced voltage into the flyback primary and secondary windings.

In the horizontal output stage, the rate of change of current, and thus the rate of collapse of the magnetic field in the flyback primary, is slowed and controlled with timing components (Ct). If the rate of the collapsing field was not slowed it would produce an induced voltage spike of several thousand volts across the flyback primary. The spikes would exceed the breakdown rating of the horizontal output transistor and produce excessive high voltages to all flyback windings. (*Figure 6-5.*)

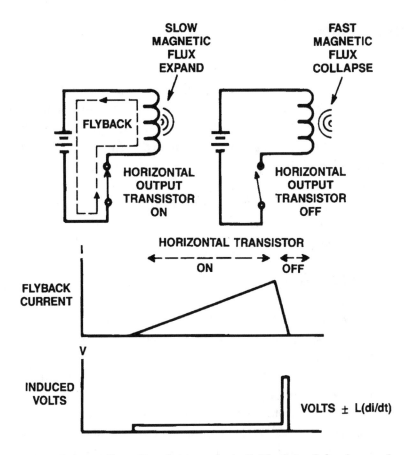

Figure 6-5. *The rapidly collapsing magnetic field of the flyback transformer, when the horizontal output transistor is switched off, produces a high voltage pulse.*

Despite the inductive "kick" action to produce high voltage pulses, the flyback transformer works similar to other transformers in transferring energy to its secondary windings. If the secondary circuits were opened, the inductive flyback transformer would return the energy stored in the magnetic field back to the primary circuit. When secondary circuits draw power (volts X current), the transformer action transfers power from the resonant primary circuit to the transformer secondaries. Some problems, such as a shorted flyback secondary circuit or flyback transformer (shorted turn) can cause excessive circuit power losses.

Retrace Timing Capacitor (Ct)

The retrace timing capacitor plays an important role in the operation of the horizontal output stage. The capacitor is sometimes referred to as a "safety" capacitor because it acts to hold down the high voltage. It does this by slowing the change in current through the flyback, and thus the collapse of the flyback's magnetic field which determines the amplitude of the flyback pulse. If the capacitor reduces in value or opens, the flyback voltage pulses would rise several thousand volts. To minimize the danger, smaller capacitors are sometimes connected in parallel. Manufacturers have added safety shutdown circuits to disable the drive or B+ to the horizontal output stage in the event of excessive high voltages.

Damper Diode (D1)

Like the horizontal output transistor, the damper diode serves as a switch. The damper diode completes the circuit path for the flyback primary and yoke currents during a portion of the resonant cycle in the horizontal output stage. Therefore, the damper diode must pass several amperes of current and switch at high horizontal frequencies. If the damper diode opens, the horizontal output transistor is forced to operate in reverse breakdown. This causes the horizontal output circuit to be inefficient, and causes the horizontal output transistor to heat and eventually fail.

Horizontal Yoke and Yoke Series Capacitor

The horizontal yoke is responsible for deflecting the CRT's electron beam. A sawtooth current in the horizontal yoke moves the CRT electron beam continually from left to right across the face of the CRT. Capacitor Cs is placed in series with the yoke to develop the resonant timing and prevent DC current from flowing. DC current to the yoke would cause improper picture centering. Capacitor Cs further shapes the sawtooth rise in current to match the slight curvature of the CRT. Because the yoke and its series capacitor are part of the horizontal output stage, they influence the resonant timing of the horizontal output stage.

Understanding the Horizontal Output Stage Operation

Now let's put the components together and see how the whole circuit operates. We will analyze the output stage in two parts according to the major functions it performs:

(a) Flyback primary current and retrace time.
(b) Horizontal deflection.

The first function, flyback primary current and retrace time, is responsible for producing CRT high voltage, focus and scan-derived supplies. The second function, as its name implies, deals with deflecting the electron beam. Although these functions interact and are part of the same circuit, discussing them separately will help you to better understand the operation of the output stage.

Flyback Primary Current and Retrace Time

Understanding the alternating sawtooth current paths of the flyback transformer primary helps to explain how the horizontal output circuit functions, and how the retrace timing capacitor affects the operation of the circuit. *Figure 6-6* breaks down the flyback action and current paths into four separate periods, beginning with the conduction time of the horizontal output transistor.

When the horizontal output transistor is on (*Figure 6-6A*), primary current rises in the flyback primary. Current is sourced from the B+ power supply. It is during this period that all the power needed by the stage and flyback secondaries is delivered to the circuit. The inductive current buildup continues until the transistor is turned off.

During the next three periods the magnetic energy stored in the flyback is exchanged with the retrace timing capacitor. The resulting current alternates through the flyback primary transferring power to the secondary load circuits.

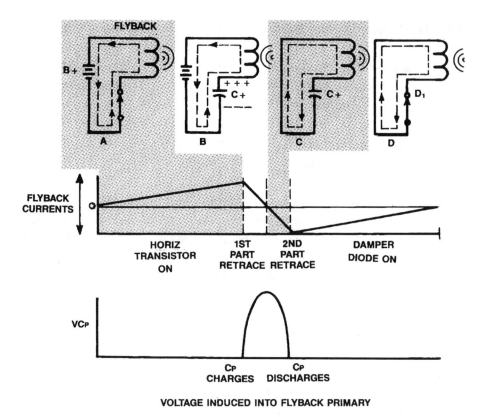

Figure 6-6. *Four equivalent circuits showing the alternating flyback current, current waveform and flyback voltage pulse for one horizontal line interval beginning with the "on" time of the horizontal output transistor.*

Immediately after the horizontal output transistor turns off, the magnetic field of the flyback begins collapsing. This is the beginning of the retrace time which corresponds to the start of horizontal sync. When the horizontal output transistor is switched off, the retrace timing capacitor is effectively placed in parallel with the flyback primary forming a resonant circuit, as shown in *Figures 6-6B* and *6-6C*. The resonant circuit timing is determined by the retrace capacitor and the equivalent flyback inductance. Timing is also slightly influenced by the yoke components in parallel with Ct.

With the horizontal output transistor off, the collapsing magnetic field of the flyback produces current flow through the low impedance of the B+ supply filter capacitors, charging Ct. The rise in voltage across Ct is the flyback pulse formed at the collector of the horizontal output transistor. This is the only voltage waveform available to help analyze operation of the horizontal output stage.

When the flyback's magnetic field is collapsed, Ct begins to discharge, reversing the direction of current through the flyback primary. This completes the second part of retrace time and is the falling portion of the flyback voltage pulse viewed at the collector of the horizontal output transistor. Properly operating the horizontal output stages of television receivers produce a total retrace period (flyback pulse duration) lasting anywhere from 11.3S to 15.9S. Computer monitors may have much shorter retrace times under normal operation depending upon their horizontal scan rates.

When Ct has completely discharged, the magnetic field begins to collapse inducing a voltage with a polarity that forward biases the damper diode D1. (*Figure 6-6D.*) The damper diode serves as a switch to allow magnetic energy in the flyback and yoke to decay at the same rate as when the horizontal output transistor was on. When the damper diode turns on, the circuit becomes highly inductive once again producing a slowly changing current in the flyback primary. Before the current fully decays, approximately 18S later in a television receiver, the horizontal output transistor is once again turned on and the cycle repeats.

Understanding Horizontal Yoke Deflection

The second major function of the horizontal output stage is to provide deflection current. Most horizontal output stages now use a direct method of providing yoke current. With this arrangement the horizontal output transistor's collector current splits two ways between the flyback and yoke. The flyback and yoke currents utilize a common damper diode and retrace timing capacitor.

Figure 6-7 breaks down the yoke deflection current into four separate events to gain a better understanding of how deflection current and timing are achieved. When the horizontal output transistor is switched on (*Figure 6-7A*), the bottom of the yoke series capacitor, Cs, is connected to the top of the yoke. Capacitor Cs is fully charged at this time and begins to discharge current through the horizontal output transistor. The resulting current produces an expanding magnetic field in the yoke which moves the electron beam from the center of the CRT to the right.

When the horizontal output transistor opens, the retrace timing capacitor is added to the circuit. (*Figures 6-7B* and *6-7C.*) This increases the resonant frequency, producing a rapid collapse of the yoke's magnetic field. This is the beginning of retrace time and the CRT beam is snapped from the right side back to the center. The induced voltage causes current to flow, returning the energy stored in the yoke to capacitors Ct and Cs. It is at this time that the retrace timing capacitor is replenished with charging current from the flyback transformer. It then serves as the current source for the yoke current.

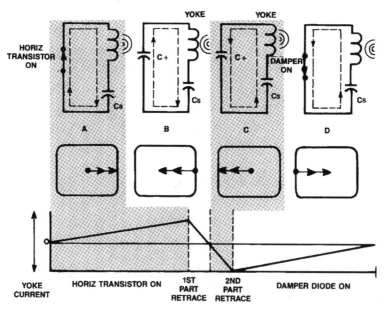

Figure 6-7. Four equivalent circuits showing the alternating flyback and yoke currents and relative position of the electron beam on the face of the CRT.

During the second part of retrace, Ct and Cs discharge and force current flow in the opposite direction, as shown in *Figure 6-7C*. The timing is identical to the first part of retrace, and the CRT beam is moved quickly from center to the left side of the screen.

When capacitors Ct and Cs are fully discharged, the yoke's magnetic field begins to collapse. (*Figure 6-7D*). The induced voltage forward biases the damper diode into conduction. The timing of the circuit is now determined by the yoke and capacitor Cs, and agrees with the timing during the right trace time. The yoke's collapsing magnetic field produces current through the damper diode, returning energy to the circuit charging Cs. The point at which the damper diode conduction stops is determined by the decay of the yokes magnetic field. In order to avoid non-linearities in the center of the raster, the timing must coincide with the point at which the horizontal output transistor yoke conduction begins.

To simplify our explanation of the horizontal output stage, we have analyzed the flyback and yoke operations separately. It should be noted that these circuits are not independent of each other. The flyback current is transferred to the yoke by the retrace timing capacitor Ct. The yoke and flyback currents share the conduction time of the horizontal output transistor and damper diode. They further share the retrace timing capacitor. Because of this interaction, most problems in the horizontal output circuits alter both the flyback and yoke currents.

Chapter Seven
Troubleshooting Secondary Voltage Circuits
By Homer L. Davidson

Determining the cause of chassis shutdown or overloaded power supply circuits, or isolating defective components in the secondary circuits of the horizontal output transformer, can consume a great deal of a technician's valuable servicing time. Overloaded circuits in any of the scan-derived voltage sources can cause chassis shutdown, which frequently makes the job of diagnosing a problem more difficult.

A shorted or leaky component in any of the circuits in a TV set can result in failure of another circuit. A defective silicon diode rectifier or electrolytic capacitor in any of the voltage sources supplied from a secondary winding of the horizontal output transformer can cause shutdown or improper circuit operation.

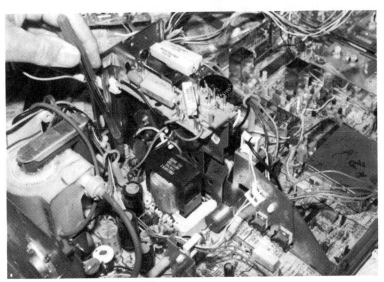

Figure 7-1. *The secondary voltage sources are derived from several windings upon the IHVT flyback.*

The secondary winding of the horizontal output transformer provides power to the horizontal output and driver circuits. In many sets, the low voltage power supply circuits may operate off of a secondary winding of the flyback as well. (*Figure 7-1*)

These various supply voltage circuits may be derived from separate transformer windings, and may be regulated and filtered by silicon diodes and electrolytic filter capacitors. Transistor or zener diode voltage regulation, or a combination of transistor and zener diode regulation, can be used in the critical voltage circuits. In some cases regulation of these voltage sources is achieved using only zener diodes.

The Various Voltage Sources

In many sets the low-voltage power supply circuit consists of a bridge rectifier network, a high voltage filter capacitor, and a DC voltage fed to the horizontal output transistor and driver circuits. The voltage source feeding these two circuits may be raw DC taken directly from the DC power supply, or it may be regulated by an IC voltage regulator. Usually the voltage supplied to the horizontal output transistor circuits is the highest DC voltage found in the set, except for boost voltage.

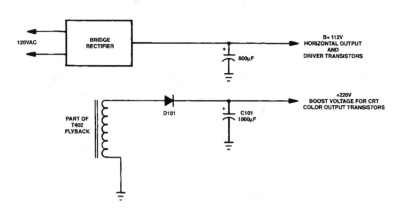

Figure 7-2. The boost voltage (+220V) in this circuit was taken from a secondary winding of the horizontal output transformer T402.

A typical flyback transformer secondary circuit supplies many different voltage sources. In the circuit shown in *Figure 7-2*, a 220V boost voltage feeds the CRT, color outputs and HV shutdown return circuits. A separate secondary winding supplies AC voltage to a silicon diode and a 47μF, 180V electrolytic capacitor. Note that the working voltage of the filter capacitor is far higher than those found in the rest of the DC sources.

Another secondary winding supplies flyback voltage to a silicon rectifier. The output of this source is used to provide several different voltages via regulation employing transistors and zener diodes. The output is also filtered using electrolytic capacitors. Most secondary circuits employ silicon diodes for half-wave rectification. In Howard W. Sams' PHOTO-FACT®, the various voltage sources are identified by number, and can easily be located on the main schematic by the number. (*Figure 7-3.*)

If, for example, you are servicing a TV set with a good picture but no sound, you would check the secondary power supply list and see what number circuit supplies voltage to the audio output circuits. You would then locate that number on the schematic to locate the point where the output of the supply appears.

The next step would be to check the voltage at that point. If the voltage is incorrect, the problem might be a defect in the supply, or a defect in the circuit that is supplied by that voltage that is causing an overload.

If the voltage is correct, the cause of the problem is not the supply.

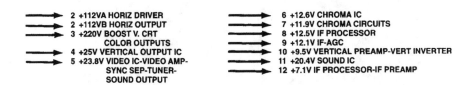

Figure 7-3. The voltage sources shown here numbered, are powered by the flyback secondary. These sources provide power to most of the circuits in this set.

Keep in mind that when one voltage source is missing, several different stages or circuits may be inoperative.

Older Flyback Circuits

In older horizontal output transformer circuits, very high HV, focus, screen and boost voltages were found in the secondary circuits. In some TV chassis, high voltage was developed from the large flyback winding with a tripler unit supplying HV to the CRT.

The focus and screen voltages are developed from a high resistance network supplied by the high voltage source. Another secondary winding, or a tap off of the bottom leg of the horizontal output transformer, supplied boost voltage to the picture tube circuits. (*Figure 7-4.*)

In later model TV sets, the integrated high voltage transformer (IHVT) was introduced. This unit comes with HV diodes and capacitors mounted inside the flyback to develop high voltage for the picture tube. The high

Figure 7-4. Older flybacks provided high voltage, screen and focus voltages from their secondary windings.

voltage resistor network from the HV output supplies both focus and screen voltages for the CRT. Still later, several different secondary windings were added to the horizontal output transformer to supply voltages to other circuits in the TV chassis.

Overloaded Circuits

Today, the IHVT transformer provides many different voltage sources derived from the secondary winding. These coils are wound on the same transformer core as the high voltage winding. If a chassis starts up and then shuts down, a likely cause is a leaky silicon diode in the secondary circuits.

Sometimes by carefully observing the screen as you turn the set on you can determine what circuit the trouble exists in. An overload in the secondary voltage circuits or in a circuit that is supplied by the secondary can cause the chassis to shut down at once.

For example, if a 25V source feeds the vertical output IC, a leaky output transistor or IC can cause an overload on that 25V source. A direct short or low resistance may damage the connecting silicon diode and isolation resistor before chassis shutdown occurs. Often, the chassis will shut down if a component in the vertical circuit is causing an overload.

In the circuit shown in *Figure 7-5*, if capacitor C322 (1000µF) had a direct short, D522 might become damaged, or the chassis might shut down before D522 incurs any damage. A leaky C335 (1000µF) electrolytic capacitor in the output yoke return circuit may delay the shut down of the chassis and cause a horizontal white line on the raster.

The sound was weak and distorted in a Panasonic AGP159 chassis. The picture was normal. The voltage at pin 9 (VCC) of the sound output IC, IC201 was only 5.4V. The voltages on both sides of R210 were low. I traced the low voltage source back to the flyback secondary 15V source. Diode D507 and capacitor C516 checked out normal. I concluded that

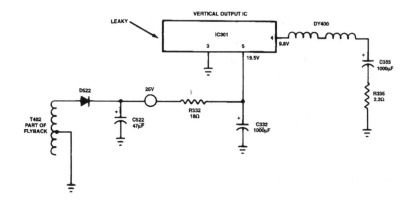

Figure 7-5. *A leaky vertical output IC, IC301, may load down the 25V secondary source, causing shutdown.*

IC201 was leaky. All of these tests were made by monitoring voltages during turn-on and shutdown operations.

I unsoldered one end of R210 to isolate this supply from the circuit it supplies, to see if the voltage source was the problem or if the IC was defective. When I again applied power to the set, the secondary 15V voltage source measured 15V (Figure 6). A resistance measurement from pin 9 to ground verified that IC201 was leaky. IC201 was replaced with an ECG1789 universal IC replacement.

Chassis Shutdown

A low resistance leakage loading down one of the voltage sources powered by the flyback secondary may cause chassis shutdown in sets that have the latest IHVT flybacks. (*Figure 7-6.*) The chassis will shut down if a horizontal deflection IC that derives its power from the flyback secondary becomes faulty.

The same type of horizontal deflection IC fault might not cause shutdown if its supply voltage comes directly from the AC-line-powered low voltage power supply. When faced with a set that is inoperative because of shutdown, try to determine what circuit might cause chassis shutdown if it becomes overloaded. Check to see what type of voltage source is feeding the horizontal deflection IC.

For instance, if you see a horizontal white line on the screen before chassis shutdown, it's a good bet that a malfunction in the vertical circuits caused the shutdown. In such a case, go directly to the vertical circuits and isolate the voltage source feeding the vertical output transistors or IC. Desolder the vertical IC voltage supply pin to isolate the supply from the IC. Likewise disconnect the isolation resistor between output transistors and the voltage source.

If the chassis starts up and remains operating (or shuts down after a few minutes), check for a leaky diode in the secondary voltage source. Refer to the schematic to locate the isolation resistor and filter capacitor. The resistor may show signs of overheating and burn marks. An open electrolytic filter capacitor will have very low voltage at the voltage source.

Figure 7-6. *Chassis shutdown may occur when one of the circuits supplied by the secondary of the flyback transformer becomes leaky and loads down the supply.*

The causes of shutdown are difficult to locate if the horizontal deflection IC circuits are fed from the secondary voltage source. The horizontal circuits must operate in order to generate the secondary flyback voltage sources. When the set goes into shutdown, you must determine if the secondary voltage sources are causing the shutdown or if the problem is caused by defective horizontal circuits.

The horizontal oscillator, countdown circuits and horizontal deflection IC can be checked by supplying a voltage from an external power supply or batteries to the horizontal circuits. With the external power source connected, use the oscilloscope to see if there is a horizontal output waveform at the deflection IC. Proceed to the rest of the horizontal circuits to determine if they are working.

Secondary Current Tests

If the set has shut down and you don't know if the problem is a faulty voltage source that derives its power from the flyback secondary, or a faulty circuit that's loading down the supply, disconnect one end of the silicon diode that provides voltage to the secondary voltage source. Disconnect only one end of the diode and check to see if the set will stay on.

All secondary voltage sources can be isolated using this method. Of course, if you disconnect the voltage source that feeds the horizontal deflection circuits, the chassis will remain in shutdown mode. Remember this same voltage source may supply voltage to several different circuits. By removing one end of each silicon diode in turn, you may be able to identify the overloaded circuit. Determine which components are causing the defective circuit using resistance and transistor-diode tests.

Another method of determining if there is an overload is to insert a current meter in series with the voltage source or silicon diode. Disconnect the positive end of the diode rectifier and insert the milliammeter. Some schematics have the total current draw marked on the various voltage sources. If the current is high on any given circuit, check to see if a faulty component is causing an overload. (*Figure 7-7.*) If the total current was specified to be 20mA, and you measured 30mA or more (for example), suspect a leaky component in the circuit supplied by that voltage source.

Voltage and Resistance Tests

Voltage measurements in the secondary circuits can determine if a component in the voltage source is faulty, or if circuits external to the power supply are leaky or shorted. Extremely low voltages may indicate an overloaded circuit or an open filter capacitor. A leaky filter capacitor in the voltage source can shut the chassis down. Leaky transistors or zener diode regulators can cause chassis shutdown as well.

Resistance measurements across the low-voltage sources may turn up a leaky component or voltage source. If voltage and resistance measurements at the terminals of a voltage source reveal low voltage and resistance to common ground, disconnect one end of the diode to determine if the voltage source is defective or if the connected circuits are defective. Monitor the suspected voltage source when the problem is chassis shut-

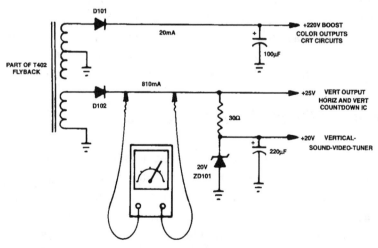

Figure 7-7. You can check to see if a circuit is causing an overload by disconnecting the positive terminal of the silicon diode that's used as a rectifier for that source, and inserting a current meter in series with the output voltage source.

down. Note if the correct voltage is there before shutdown. Most likely the voltage source is normal if correct voltage is found before shutdown. Suspect leaky components in the voltage source circuits when the voltage at the source terminals remains at 0V.

A diode test of silicon diodes, regulator transistors and zener diodes may turn up the defective component. Measure the resistance across the secondary electrolytic capacitor to see if there is leakage or low resistance. When low resistance is found across the secondary voltage source, disconnect the external circuit or check that circuit for leaky or shorted components.

In an RCA CTC146 chassis, the TV set would come up and immediately shut down. The 140V, 160V, and 185V sources were fairly normal. The standby voltage was very low. It should be around 12V. According to the schematic, the standby regulator (Q3107) had sufficient input voltage to the collector terminal. At first I suspected that the regulator transistor was open, but it tested normal in and out of the circuit.

Resistance across CR3105 and CR3104 in the base circuit of the 12V regulator measured below 27Ω. Further checking of these components out of circuit revealed that CR3104 had a direct short across its terminals and CR3105 was leaky. (*Figure 7-8.*) Replacing both the 6.8V and 5.6V zener diode regulators solved the chassis shutdown symptom.

Secondary Open Filter Capacitors

Very low secondary voltage sources can be caused by a leaky silicon diode, burned isolation resistor, or open filter capacitor. Check the suspected diode and resistor with a resistance test. The dried-up or open filter capacitor in the secondary circuits may be a little more difficult to locate. You could try an in-circuit capacitor test, but this measurement is fruitless, since the surrounding resistances in the output voltage source are low.

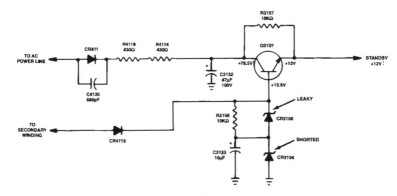

Figure 7-8. Shorted CR3104 and leaky CR3105 caused the RCA CTC146 chassis to shut down at once.

When the voltage of a source is low, and the chassis is not in shutdown mode, turn off the set, disconnect the line cord and shunt a known good electrolytic capacitor with correct working voltage across the suspected one. Apply power to the set and check to see if the voltage returns to normal. If so, remove the open capacitor and test it out of the circuit. If this test confirms that it's bad, replace it.

Although you can remove the suspected capacitor and test it out of the circuit, this test may indicate that the capacitor is normal, even if there's a broken internal connection. The broken connection may be temporarily restored by the jostling experienced by the capacitor as it's being removed from the board. Shunting the capacitor in the circuit is much quicker and accurate, because it doesn't disturb the condition of the electrolytic capacitor.

Identifying Secondary Voltage Sources

It can be very difficult to service secondary voltage sources without a schematic. In some sets, the voltages of the sources are stamped on the printed circuit board, making things a little easier. Howard W. Sams' PHOTOFACT® lists the various voltage sources by number. By locating the correct silicon diode in the secondary voltage sources, you can locate the diode on a chassis layout chart.

Critical voltage sources can be measured or monitored from the positive terminal of the silicon diode. Look for these secondary components on the printed circuit board near the flyback transformer. If the schematic is not available, it may be possible to trace out the various secondary windings from the transformer to the corresponding silicon rectifier, and create a rough schematic of your own.

Secondary Voltage Regulators

You may find several transistors and zener diode regulators in the various secondary low voltage circuits. In the 20V source of the circuit of *Figure 7-9*, a 19V zener diode regulates the 20V source. This voltage source is fed to the first and second video amp, sync, tuner, and sound output circuits. The 11.9V source has a transistor as voltage regulator (Q601). The 11.9V source feeds the chroma IC circuits. The 12V source is regulated with a 12V zener diode. Do not overlook the possibility that the problem is an open or leaky transistor or zener diode regulator if the voltage from a particular secondary voltage source is low.

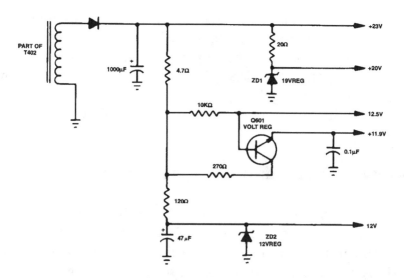

Figure 7-9. Check for leaky or open voltage regulator transistors and zener diodes in the defective scan-derived voltage sources.

Conclusion

When servicing voltage supply circuits that derive their power from a horizontal output transformer, determine if the horizontal oscillator or deflection IC is powered directly from the low-voltage power supply or if they're powered by the secondary voltage source. Chassis shutdown problems are easier to service if the supply voltage for the deflection IC comes from the low-voltage power supply.

The horizontal circuits must be fully operational if the supply voltage of the deflection IC is taken from the secondary voltage sources. Check for leaky components in the secondary voltage sources and overloaded components in the connecting TV circuits.

Chapter Eight

TV Power Supply Troubleshooting
By Brian E. Jackson

A power supply is an electronic circuit that performs some type of power conversion. Modern TV sets use switching power supplies to furnish DC power to most of the circuits. However, these sets also contain one or more linear power supplies.

These linear power supplies are used for operating the switch-mode DC/DC converter and for powering standby circuits such as the remote-control receiver and microprocessor. This article will discuss the theory of operation and the troubleshooting of the linear power supply for a TV receiver.

DC-to-DC Converter

A DC-to-DC (DC/DC) converter converts DC of one voltage into a DC voltage of another value. For example, a DC/DC converter might be used to convert +5V into +12V and -12V, as shown in *Figure 8-1*. A +5V supply is common in many pieces of electronic equipment. Instead of building in additional +12V and -12V supplies, these voltages can be derived from the +5V supply by using a DC/DC converter.

A DC/DC converter uses pulse and switching circuits to achieve the voltage translation. We refer to this as a switch-mode supply because the output of the supply is generated by using transistor switches to convert this DC into a high-frequency AC signal.

This AC is then passed through a transformer to step-up or step-down the voltage to the desired level. Then another rectifier converts the square-wave pulses into pulsating DC, which is then filtered into a constant DC by a capacitor filter.

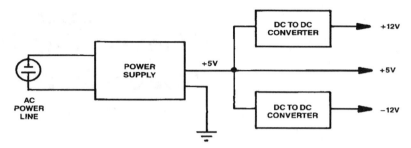

Figure 8-1. A DC-to-DC converter converts DC of one voltage to DC of a different voltage.

Switch-mode power supplies have become popular because the use of high-frequency AC as the input to the voltage translation transformer allows much smaller, lighter and less-expensive (less copper) transformers to be used.

DC/DC Converter Components

Figure 8-2 shows the basic components of a DC/DC converter. The DC supply is any normal driven power supply. Its voltage is passed through Q1, a high power switching transistor.

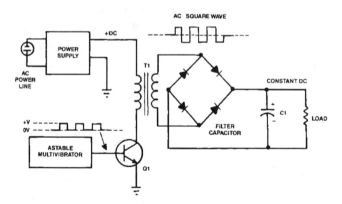

Figure 8-2. This DC-to-DC circuit uses an astable multivibrator to convert the DC output of a power supply to an AC square wave. The AC square wave is then transformed in voltage by the transformer, rectified and filtered. The output is a DC voltage that is different in level from the original power supply output voltage.

This transistor is driven by an astable multivibrator. When the transistor turns on, current flows through the primary winding of transformer T1, current flows, and a magnetic field builds up. When Q1 switches off, the magnetic field collapses.

The changing magnetic field induces a voltage into the secondary of T1. The transformer is set to deliver the required voltage across the secondary. These pulses are rectified by the bridge rectifier into a pulsating DC. The capacitor filters this into a constant DC voltage.

Horizontal Output Supply

As you saw in the previous section on the DC/DC converter, a main source of DC is required to operate the oscillator, power switches and regulators that generate the AC pulse signals that are ultimately rectified into the DC voltages used to power most TV circuits. In modern TV sets, the horizontal output stage is the base of this DC/DC conversion.

Figure 8-3 shows the supply that is used to power the horizontal sweep and output circuits. Most manufacturers use this circuit or some variation of it.

The AC from the line cord is first applied through a fuse to two inductors, L1 and L2. These inductors, along with C1, form a low-pass filter that serves two purposes. First, it keeps high-frequency noises on the power line from reaching the other circuits in the set. Second, it prevents high-frequency noise and signals developed inside the set from getting back into the power line.

Notice that this circuit does not use a power transformer. Some power supply circuits use a transformer to isolate the AC from the rectifier, and to either step-up or step-down the AC line voltage to the desired level before rectification. In this circuit, the AC line voltage is applied to a bridge rectifier circuit. Capacitors C2 through C5 suppress any high-frequency noise. The DC across filter capacitor C6 is about 150V unregulated.

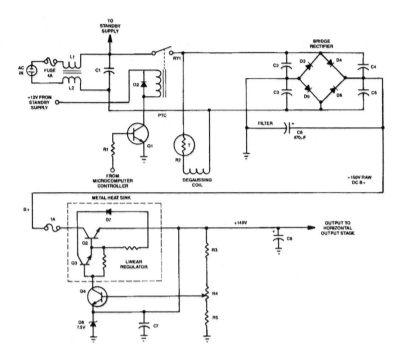

Figure 8-3. Most TV manufacturers use this circuit, or some variation of it, to power the sweep and output circuits.

This voltage is often referred to as "raw" DC, because it is unregulated. This voltage is frequently called B+. Some sets apply this raw DC voltage directly to the horizontal output stage. Usually, however, some kind of series regulator circuit is used between the filter output and the horizontal stages, as shown in *Figure 8-3*.

Note that the B+ is applied through a separate fuse to the series linear regulator. R4 allows you to set the output voltage to the precise value desired. Diode D8 is used as the reference to which the output voltage sample is compared. Q4 in *Figure 8-3* generates an error signal that drives the base of the series pass element, which is a Darlington transistor. Physically, this device looks like a regular power transistor and is usually mounted on a heat sink to help dissipate the heat it generates.

The two transistors, Q2 and Q3, are connected internally along with the bias resistors and the reverse protection diode, D7. The output is a constant 140V DC, which serves as the main power source for the horizontal circuits and all of the scan derived voltages. C8 is a small electrolytic that provides final filtering.

Also, note in *Figure 8-3* that a degaussing circuit is shown. Most degaussing coils use a positive temperature coefficient (PTC) thermistor. When the circuit is first turned on, R2 has a very low resistance so maximum current flows in the degaussing coil and the resulting magnetic field does the degaussing. The thermistor heats up fast and its resistance increases to a very high value, effectively reducing the current flow to a very low level.

Standby Supply

Modern TV receivers usually produce standby voltage for remote control operation. Standby voltage is available even when the set is turned off. In *Figure 8-3*, the relay contact, RY1, connected in series with one side of the AC line, is driven by standby voltage. When you turn the set on using the remote control, a coded signal is sent to a remote receiver which is then interpreted by a microprocessor.

The remote receiver and microprocessor are also powered by standby voltage. The microprocessor generates a logic signal which is sent to the base of Q1 through R1. Q1 turns on and causes current to flow in the relay coil. This creates a magnetic field that closes the relay contact. This AC is applied to the bridge rectifier and the main supply turns on.

A standby supply is a power supply used to power standby circuits. The main standby circuits are the remote control receiver and the microprocessor that interprets inputs from the remote and issues control signals to the rest of the set. The remote receiver must, of course, be turned on all the time so it can respond to a signal from the remote control.

The standby supply is usually a half-wave rectifier diode, D1, connected to the AC input right after the input filter. (See *Figure 8-4*.) The take-off point is shown in *Figure 8-3*. A resistor, R1, is used to drop the unused voltage produced by the line. C2 filters the 60Hz pulses from the rectifier into a steady DC. Next, this DC is applied to two series pass transistors. The two output voltages are +5 and +12 volts. The +5V powers the single-chip microcomputer and any auxiliary circuits. The +12V powers the remote receiver circuits.

Power Supply Troubleshooting

A multimeter with a high voltage attachment and an oscilloscope are usually all you need to troubleshoot various power supplies, including switch-mode power supply circuits.

Common Symptoms

Most TV sets made since 1982 use some type of switching power supply arrangement. Because this circuit uses feedback, the usual symptom when something goes wrong with the power supply is a completely dead set. However, it is also quite possible to have a power supply problem with the set partially functional. For example, if a rectifier diode in one of the circuits that operates the tuner is open, the entire power supply wouldn't shut down.

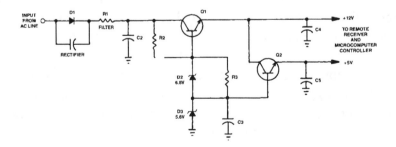

Figure 8-4. The standby power supply in a TV set is usually a half-wave rectifier diode (D1 in this circuit) connected to the AC input right after the input filter.

A picture with no audio suggests the possibility of a partial power supply failure: you might have a bad power supply circuit for the audio section. On the other hand, the audio circuits themselves could be bad. In any case, you should always check the DC voltage supplying a circuit first before you try troubleshooting anything else. If the voltage is present, you troubleshoot the inoperative circuit. If the voltage is absent, you check the power supply to see why.

Other symptoms that suggest a bad power supply include distorted audio or hum, a shrunken or dim picture, or lack of horizontal or vertical sync. These can sometimes be caused by reduced power supply voltages or an increase in ripple. This means that the power supply is working, but not as it should.

Defective Components

One thing that will help you quickly find the faulty part is the knowledge that some parts are more likely to fail than others. After years of keeping track of TV set failures, most manufacturers have found that statistically you can almost predict what the problem will be. And these problems are the same for any manufacturer.

Listed are the parts that are most likely to fail, given in the order from the most likely to the least likely:

(a) Bad power transistors (horizontal output or driver, series pass regulator, etc.).
(b) Bad rectifier diodes (usually shorted).
(c) Defective IC voltage regulators.
(d) Open power resistors (wattage rating of 2W or more).
(e) Defective flyback transformer.
(f) Bad filter capacitor (electrolytic).
(g) Broken wire or bad connector.
(h) Bad off/on switch.

In many cases, you will experience two or more of these problems at the same time. For example, if a filter capacitor shorts, it usually takes out several other components, such as a power transistor, rectifier diode and power resistor. Don't stop your troubleshooting when you have located one bad part; look at other parts associated with the bad one.

The following is a recommended troubleshooting procedure:

Verify the Problem

The first thing you should do is to try to validate the customer's complaint and attempt to duplicate the symptoms. You really must see the problem yourself just to be sure you are getting all of the data you will need to figure out where the problem is. The customer's complaint may contain some useful information, but always check the symptoms yourself.

Perform a Visual Analysis

You should first check the most obvious problems. Is the line cord to the set really plugged in? If it is, is there AC at the outlet? A fuse or circuit breaker may have opened, shutting off power to the outlet.

When you are satisfied that you are getting AC power to the set, you can continue. Next, you should open up the set and look around inside. Refer to the manufacturer's service literature to find your way around inside the set. Locate the power supply circuit boards and any related assemblies, such as the flyback transformer.

Now, visually inspect these items. Look for broken or burned parts. Check for broken wires and dirty or unseated connectors. Look for melted wax that has flowed out of a transformer, indicating that this part has over-heated. Of course, if you locate a defective part, then you are on the right track. Remember, though, that the damage to a burned out or defective part may have been caused by the failure of another part.

Check Circuit Breakers and Fuses

All power supplies contain some type of circuit breaker that protects the circuitry if something goes wrong. A fuse, for example, will blow if the current demand on the circuit is greater than usual. A shorted part or circuit usually causes this. You can usually tell if a fuse is good just by looking at it.

Most TV fuses are small glass cylinders containing a thin wire. If the wire is still there, most likely the fuse is still good. Keep in mind, though, that on rare occasions there may be a bad connection within the fuse, and it will not pass current even though it appears to be intact.

If the wire is missing, or the fuse glass housing is blackened, then the fuse is bad. If you aren't sure, check the continuity with your ohmmeter.

Replacing fuses can be tricky, particularly if they are soldered in place. In any case, just be sure you replace the fuse with the exact value of the original.

Perform a Voltage Analysis

Voltage analysis is a step-by-step procedure for isolating a defective part by tracing the voltages in a particular section of the receiver with a multimeter. This will usually lead you to the bad circuit; then you can make some component tests with your ohmmeter to pinpoint the bad part.

A good starting point for a voltage analysis is to measure the regulated output voltage of the linear supply. It should be around 140V. Remember to refer to the service manual to locate parts inside the receiver.

In addition, the manufacturer will often provide troubleshooting charts to help you locate the problem. Refer to these charts often and apply deductive reasoning to guide you in the repair.

Check for High Voltage Problems

High voltage problems usually require some special troubleshooting procedures. Generally, the procedures for troubleshooting any scan-derived supply apply. If the horizontal output transistor (HOT) fails, the HV will be lost. If the horizontal oscillator, countdown IC, or driver fail, no pulses will reach the HOT, so there will be no scan-derived voltage, HV included. Verify these circuits first.

To verify the absence of high voltage, measure it using an HV probe or with an HV meter. This typically comes in the form of a large, insulated probe that contains a small analog meter which reads directly in KV. These are not too expensive, and they are handy if you plan to do a lot of TV service work. They are also safe, provided you read the manufacturer's suggested HV value, and use it only in the manner recommended.

Make the Repair

Once you have identified and removed the bad component, check it with the ohmmeter to be sure. However, to make sure that the identified component is the only defective one, substitute a good component for the defective one in the circuit, and check to see if normal operation is restored. If normal operation is restored, you've solved the problem.

Glossary

Some of the terms used in this article may not be familiar to all readers. The following is a glossary of the most important terms used.

Astable Multivibrator: An astable circuit is a form of oscillator. The word astable means unstable. An astable multivibrator consists of two tubes or transistors arranged in such a way that the output of one is fed directly to the input of the other. The astable multivibrator is frequently used as an audio oscillator, but is not often seen in RF applications because it is extremely rich in harmonic products.

B Plus: The term B plus (B+) refers to the positive high voltage dc supply used for the operation of a vacuum tube circuit. Since tubes are becoming less and less common in modern electronics, the expression B+ is heard less often today than a few years ago, but sometimes is applied to the voltage source in transistors. B+ voltages are frequently high enough to be dangerous.

Bridge Rectifier: A bridge rectifier is a form of full-wave rectifier circuit, consisting of four diodes. Bridge rectifier circuits are commonly used in modern solid-state power supplies. Some integrated circuits are built especially for use as bridge rectifiers; they contain four semiconductor diodes in a bridge configuration, enclosed in a single package.

Capacitor: A capacitor is a device designed for providing a known amount of capacitance in a circuit. Capacitors are available in values ranging from less than 1pF to hundreds or thousands of F. It is rare to see a capacitor with a value approaching 1F. Capacitors are also specified according to their voltage-handling capability; these ratings must be carefully observed, to be certain that the potential across a capacitor is not greater than the rated value. If excessive voltage, either from a signal or from a source of direct current, is applied to a capacitor, permanent damage can result.

Continuity: In an electrical circuit, continuity is the existence of a closed circuit, allowing the flow of current. A simple device called a continuity tester may be used to check for circuit continuity. The continuity tester consists of a power source and an indicating device, such as a battery and a lamp.

Converter: Any device that converts frequency, voltage or current, or computer data from one form to another is called a converter.

Darlington Amplifier: A Darlington amplifier, or Darlington pair, is a form of compound connection between two bipolar transistors. In the Darlington amplifier, the collectors of the transistors are connected together. The input is supplied to the base of the first transistor.

The emitter of the first transistor is connected directly to the base of the second transistor. The emitter of the second transistor serves as the emitter for the pair. The output is generally taken from both collectors.

The amplification of the Darlington pair is equal to the product of the amplification factors of the individual transistors as connected in the system. This does not necessarily mean that a Darlington amplifier will produce far more gain than a single bipolar transistor in the same circuit. The impedances must be properly matched at the input and output to ensure optimum gain.

Some Darlington pairs are available in a single case. Such devices are called Darlington transistors. The Darlington amplifier is sometimes called a double emitter follower or a beta multiplier.

Degaussing: Degaussing is a procedure for demagnetizing an object. A device called a degausser or demagnetizer is used for this purpose. Sometimes the presence of a current in a nearby electrical conductor can cause an object to become magnetized.

This effect may be undesirable. Examples of devices in which degaussing is sometimes required include TV picture tubes, tape recording heads and relays. By applying an alternating current to produce an alternating magnetic field around the object, the magnetization can usually be eliminated. Another method of degaussing is the application of a steady magnetic field in opposition to the existing, unwanted field.

Diode: A diode is a vacuum-tube or semiconductor device intended to pass current in only one direction. The positive terminal of a diode is called the anode, and the negative terminal is called the cathode, under conditions of forward bias. Semiconductor diodes are used for many different purposes in electronics. They can be used as amplifiers, frequency controllers, oscillators, voltage regulators, switches, mixers, and in many other types of circuits.

Feedback: When part of the output from a circuit is returned to the input, the situation is known as feedback. Sometimes feedback is deliberately introduced into a circuit; sometimes it is not wanted.

Feedback is called positive when the signal arriving back at the input is in phase with the original input signal. Feedback is called negative (or inverse) when the signal arriving back at the input is 180 degrees out of phase with respect to the original input signal.

Positive feedback often results in oscillation, although it can enhance the gain and selectivity of an amplifier if it is not excessive. Negative feedback reduces the gain of an amplifier stage, makes oscillation less likely, and enhances linearity.

Filter Capacitor: A filter capacitor is used in a power supply to smooth out the ripples in the DC output of the rectifier circuit. Such a capacitor is usually quite large in value, ranging from a few microfarads in high-voltage, low-current power supplies to several thousand F in low-voltage, high-current power supplies. Filter capacitors are often used in conjunction with other components such as inductors and resistors.

The filter capacitor operates because it holds the charge from the output-voltage peaks of the power supply. The smaller the load resistance, the greater the amount of capacitance required in order to make this happen to a sufficient extent.

Filter capacitors in high-voltage power supplies can hold their charge even after the equipment has been shut off. Good design practice recommends that resistors of fairly large value be placed in parallel with the filter capacitors in such a supply so that the shock hazard is reduced. Before servicing any high-voltage equipment, the filter capacitors should be discharged.

Half-Wave Rectifier: A half-wave rectifier is the simplest form of rectifier circuit. It consists of nothing more than a diode in series with one line of an alternating-current power source, and a transformer (if necessary) to obtain the desired voltage.

Low Pass Filter: A low pass filter is a combination of capacitance, inductance, and/or resistance, intended to produce large amounts of attenuation above a certain frequency and little or no attenuation below that frequency. The frequency at which the transition occurs is called the cutoff frequency.

At the cutoff frequency, the attenuation is 3dB with respect to the minimum attenuation. Below the cutoff frequency, the attenuation is more than 3dB.

Scan-Derived Power Supplies: A scan-derived power supply is a DC/DC converter that uses the horizontal output stage and flyback transformer. The term scan refers to the sweep circuits used to generate the ac pulses that are rectified into DC voltages. These voltages are used to power all of the circuits in the TV set.

Step-Down Transformer: A step-down transformer is a transformer in which the output voltage is smaller than the input voltage. The primary-to-secondary turns ratio is the same as the input-to-output voltage ratio. The input-to-output-impedance ratio is equal to the square of the input-to-output voltage ratio.

Step-Up Transformer: A step-up transformer is a transformer in which the output voltage is larger than the input voltage. The input-to-output voltage ratio is equal to the primary-secondary turns ratio. The input-to-output impedance ratio is equal to the square of the input-to-output voltage ratio.

Temperature Coefficient: Many electronic components are affected by fluctuations in temperature. Resistors and capacitors, especially, tend to change value when the temperature varies over a wide range. The tendency of a component to change in value with temperature variations is known as temperature coefficient.

If the value of a component decreases as the temperature rises, that component is said to have a negative temperature coefficient. If the value increases as the temperature rises, a component has a positive temperature coefficient.

A few components exhibit relatively constant values regardless of the temperature; these devices are said to have zero temperature coefficient. The temperature coefficient is usually expressed in the percent per degree of Celsius.

Chapter Nine
A Brief Look at Color Television Receiver Circuits
By Lamarr Ritchie

A color TV receiver contains all of the circuitry of the monochrome receiver, plus the added circuits needed to demodulate and display the color portion of the picture. To display the picture in color, three video signals are derived; the original red, green and blue video signals. The color CRT has three color phosphors, each of which glows with one of the three primary colors when bombarded by electrons. These phosphors are placed on the inner surface of the CRT as either triangular groups of the three colors (mostly in older receivers), alternating rectangles of the three colors, or alternating stripes of the three colors. Regardless of the version, all color CRTs require three separate electron beams, each modulated with the video of one of the primary colors. Each also has a shadow mask placed behind the phosphors. This mask has a series of openings that allows each electron beam to strike only the correct color phosphor. The three beams must be precisely aligned to enable them to enter the opening in the mask at the correct angle and strike the correct phosphor. Stray magnetic fields could create enough error to cause the incorrect color to be displayed in parts of the picture. For this reason color CRTs require an automatic degaussing circuit to keep the CRT and mounting components demagnetized. The automatic degaussing coil is mounted around the CRT just inside the cabinet. *Figure 9-1* is a general block diagram of the circuits found in a color TV receiver. There may be other circuits, and some may differ, but this block diagram is a good starting point when it comes to developing an understanding of what the circuits in a color TV set do and how they are interconnected.

Color TV Main Stages

Here is a brief description of the main stages found in a color TV receiver. This article will often make references to "older receivers" or "mod-

---

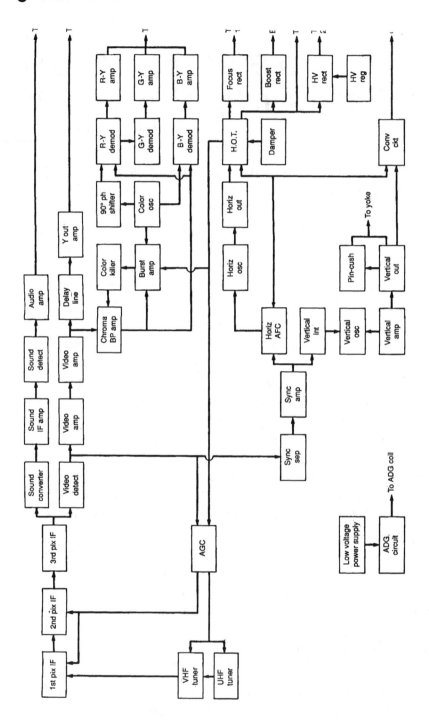

Figure 9-1. A general block diagram of the circuits found in a color TV receiver.

ern receivers." More often than not, a consumer electronics technician will be servicing a set that's several years old, rather than more recent models:

UHF Tuner - Converts the UHF channel selected down to the common video IF frequencies. Older tuners were arranged as shown on the block with the UHF tuner feeding into the VHF tuner. When UHF stations were selected, the VHF tuner's mixer (and on some, the RF amp also) became video IF amps for extra gain.

VHF Tuner - Converts the VHF channel selected down to the common IF frequencies. At the tuner output, the picture carrier is at 45.75 MHz and the sound carrier is at 41.25 MHz. Note that in later model receivers, UHF and VHF tuner circuits are combined into a single unit that is also capable of tuning to the additional channels for cable TV.

Video IF Amplifiers - These amplify the video and audio frequencies within the video IF pass band to the levels sufficient to drive the video detector and sound converter. There are typically three or four video IF amplifier stages, sometimes referred to as PICTURE IF AMPS or simply PIX IF AMPS.

Sound Converter - This stage converts the sound frequency in the video IF passband (41.25 MHz) down to the sound IF frequency of 4.5 MHz by heterodyning the picture and sound frequencies together and tuning to the difference frequency. This circuit could be simply a diode and 4.5 MHz tuned circuit.

Sound IF Amp - Also referred to as audio IF amp. This stage is responsible for amplifying the 4.5 MHz sound IF signal to a level sufficient enough to be detected by the sound detector.

Sound Detector - Also known as an audio detector. Converts the FM modulated 4.5 MHz sound IF frequency to baseband audio.

Audio Amplifier - This circuit will amplify the audio frequencies from the detector to the power levels sufficient to drive the speaker. If the TV is MTS equipped, there may be two audio amplifiers (stereo amplifiers) and the MTS decoding circuits will precede the audio amps.

Video Detector - An AM detector, usually a diode detector, that converts the picture frequencies within the video IF passband down to baseband video. Immediately following the detector will be a 4.5 MHz sound trap to remove the frequency that will result from the heterodyning of the picture and sound frequencies of the video IFs.

Video Amplifiers - These amplify the video from the detector to levels sufficient to drive other circuits.

Delay Line - This is a device that delays the Y signal (monochrome video) about 0.8 seconds to match the delay in the chroma circuits. The color signal is delayed more because more circuitry is involved in recovering the color components of the signal.

Y Output Amp - Also known as the video output amp. For the color TV versions that use it, this stage amplifies the Y signal (black/white or brightness component) to the level sufficient to drive the CRT. Most color receivers use R, G, and B video output amps with the Y signal already mixed back in with the three color-difference signals.

Chroma Bandpass Amp - This stage separates the color components from the video signal and amplifies them to a level sufficient to drive the demodulators and burst amplifier. There may be 2 or 3 chroma bandpass amps.

Burst Amplifier - The purpose of this block is to separate the color burst (color sync) from the chroma signal and amplify the burst. This signal will be used to phase lock the color oscillator.

Color Oscillator - A crystal oscillator at 3.58 MHz (actually 3.579545 MHz) that supplies the reference signal for demodulating the color. This frequency was suppressed during the modulation process.

Color Demodulators - The B-Y demodulator recovers the B-Y signal using the chroma and oscillator CW at 0 degrees. The R-Y demodulator recovers the R-Y signal using the chroma signal and the oscillator CW at 90 degrees. The 90 degree phase shifter in the block diagram shifts the oscillator phase for use by the R-Y demod. Most newer TV receivers also use a G-Y demodulator, but some use a G-Y adder circuit that derives the G-Y signal by adding together the proper amounts of the R-Y and B-Y signals. This is the type shown on the preceding block diagram. Another type, called X-Z demodulators, will be described later.

Color Difference Amplifiers - These amplify the R-Y, G-Y and B-Y signals to a level sufficient to drive the CRT. These are used only if there is a separate Y video output amp, as described above.

Automatic Gain Control, or *AGC* - This stage keeps a constant video level at the detector by closely controlling the gain of the tuner's RF amp and the video IF amps.

Sync Separator - Removes the sync pulses from the composite video signal for use by the deflection circuits.

Sync Amplifier - The sync amp amplifies the sync pulses to a level sufficient to be used by the deflection circuits. In modern receivers it may also provide pulses for the AGC circuit and digital circuits, such as channel/time display, etc.

Horizontal Oscillator - This will develop the 15.75 KHz (15,734.26 Hz to be exact) horizontal scanning frequency.

Horizontal AFC - Automatic frequency control circuit for the horizontal oscillator. Its purpose is to keep the horizontal oscillator locked to the horizontal sync pulses.

Horizontal Output Amp - Amplifies the 15.75 KHz pulses from the horizontal oscillator to a level sufficient to drive the horizontal deflection windings in the deflection yoke. It may also have to provide power for other circuits as there may be secondary windings of the horizontal output transformer that are rectified and used as DC power supplies.

Horizontal Output Transformer - This is more commonly called the flyback transformer. It may also be known as an integrated flyback transformer (IFT) or integrated high voltage transformer (IHVT) if the high voltage rectifier is built in to the transformer. Its primary purpose is to match the horizontal output stage to the horizontal deflection yoke winding, but it has many other functions. It develops the high voltage AC pulse at 15.75 KHz, which is rectified and used for the CRT's second anode. It is also used to develop keying voltage pulses and DC supply voltages for other circuits.

Damper - The yoke and flyback transformer have high inductance and stray capacitance between their windings that can cause ringing when shock excited by the abrupt change in scanning current during flyback time. The damper is simply a diode, usually connected across the horizontal output amp, that conducts during retrace (flyback) time and acts as a low impedance during this time to damp out this ringing.

High Voltage Rectifier - Rectifies the high voltage AC pulse from the flyback transformer to produce the high voltage dc needed by the CRT. A filter capacitor is not needed because the Aquadag coating on the inside and outside of the CRT act as a high voltage capacitor.

High Voltage Regulator - Keeps the high voltage constant in spite of varying beam currents (varying brightness of the picture) and line voltage changes. This is necessary because changes in the high voltage will cause misconvergence of the three beams.

Convergence Circuit - Specifically, this means the circuitry for the dynamic convergence of the three electron beams. This circuit generates the waveforms that feed the convergence yoke. It makes minor corrections to the main scanning voltage to insure that each of the three beams strike the correct place on the CRT screen at all points in the scanning process.

Only the older delta type CRTs require a complex dynamic convergence circuit. Newer in-line and Trinitron CRTs require only adjustable permanent magnets placed around the neck of the tube.

Boost Rectifier - When the horizontal output begins to supply power to the flyback, between pulses, the collapsing field of the primary winding produces a counter-counter EMF that aids, and is approximately equal to, the DC supply voltage. This may be rectified and used in places that need more than the normal dc supply voltage, such as for the CRT screen voltages. Most modern receivers do not use the boosted B+ voltage, instead using multiple secondary taps on the flyback transformer.

Focus Rectifier - Rectifies the voltage at a flyback transformer tap to produce the moderately high DC voltage for the first anode, used to focus and accelerate the electron beam. Most modern receivers do not use a focus rectifier, instead using a network of high voltage resistors to reduce the main high voltage to the correct level. Many also use a tap in a voltage tripler that is employed after a lower voltage flyback transformer.

Vertical Integrator - Usually considered as part of the vertical oscillator. A low pass filter that removes the vertical sync pulses from the composite sync to be used to synchronize the vertical oscillator.

Vertical Oscillator - This circuit generates the 60 Hz (actually 59.94 Hz) vertical scanning frequency.

Vertical Amps and Output Amp - These amplify the vertical scanning signal produced by the vertical oscillator to a level sufficient to drive the vertical windings of the deflection yoke. In older receivers a vertical output transformer was used to match the vertical output amp to the yoke. Most modern receivers do not require it because transistor amps, such as the complementary symmetry, have an impedance low enough to drive the yoke directly.

Pincushion Circuit (PIN) - Develops a parabolic waveform that compensates for the "pincushion" effect caused because the electron beams must travel a greater distance to reach the edges of the screen than they do to reach the center area of the screen. Depending on the size and deflection angle of the CRT, there may be side and top/bottom pincushion circuits. For modern receivers this might not be a discrete circuit but an integral part of the scanning circuits, if used at all.

Low Voltage Power Supply - Supplies operating dc voltages for the various circuits. Most modern receivers do not really have a low voltage power supply that is operated from the ac power line instead using a scan derived power supply. Only the voltages needed to start the horizontal circuits are derived from the AC line. Most of the other voltages would be derived from the flyback transformer, once the horizontal circuits are operating. This is the most cost effective way to generate these voltages because fewer components are needed and the higher frequency pulses are easier to filter, requiring smaller filter capacitors.

Automatic Degaussing Circuit, or *ADG* - Demagnetizes the CRT screen and metal holders each time the TV is turned on to prevent stray magnetic fields from interfering with color purity.

Color Television Receiver Circuits: Part 2
By Lamarr Ritchie

Part 1 of this article provided an overview of the circuits in a color TV receiver, including definitions of the operation of the major stages. This second part of the article will detail the operation of the video IF amps, the video amps, the sync circuits and the AGC circuits.

RF Tuners

The VHF tuner contains, at a minimum, an RF amplifier, local oscillator, and mixer stages. The local oscillator operates at a frequency that is 45.75 MHz above the visual carrier frequency. This produces video IF frequencies of: (a) Picture 45.75 MHz, and (b) Sound 41.25 MHz. Notice that in the video IF passband, the picture carrier is 4.5 MHz above the sound carrier, opposite to the way the carriers are oriented for the original RF signal. To receive an acceptable, snow-free picture, the S/N (signal to noise) ratio at the tuner input should be at least 30:1. This normally requires about 500V of signal strength in the VHF band. Signal strength is also measured in dBm. (dB relative to 1mV.) Often dBm is shortened to dB for conversational purposes; therefore, 0dB = 1000V of signal. To maintain a good quality picture, the signal level usually must be above 0dB. The output of a typical VCR or computer/game modulator is about 6dB (2000V). Most older tuners had wavetraps for the FM radio band (88 MHz to 108 MHz) but newer "cable-ready" receivers do not because these frequencies are used for cable-only channels.

Tuner Inputs

Tuners have 300Ω balanced inputs, using two terminal connectors, and/ or a 75Ω "F" type cable connector. The 300Ω balanced type is primarily for antenna connections for the reception of local signals. The balanced connections provide minimum pickup on the lead-in wires of signals that would cause ghosts and interference, and 300Ω is the most common impedance for TV antennas. For long distances, however, 75Ω coax cable is preferred, and many modern receivers have only this type of connection. For changing the connection from one type of lead-in wire to the other, a BALUN (balanced to unbalanced) transformer is used, having an "F" type cable connector on one end and two terminal connectors on the other. Each time the received channel is changed, at least four circuits in the tuner must be changed in frequency: the antenna input tuning, RF output amp tuning, mixer input tuning, and local oscillator frequency. To accomplish this, older tuners usually used one of two mechanical arrangements to change the coils for the circuits for each channel.

Mechanical Tuners

The "turret type" tuner used long rectangular turrets, each of which contained all of the coils needed to tune a particular channel. These were arranged like the old Gattling gun. As the channel was changed, a turret moved around until it was in position against a set of contacts for each coil. Usually, the local oscillator coil was in front and had a gear-driven core, driven externally by a knob to adjust the fine tuning for each channel. The other popular type of mechanical tuner was the "wafer type." This one used stacked/ganged wafer switches to change the coils. The biggest problem with these tuners was that the contacts would become dirty and/or worn, requiring frequent cleaning or adjusting.

Varactor-Diode Tuners

Modern tuners use varactor diodes to accomplish this tuning. The use of varactor diodes allows the circuits to be tuned using only a DC control voltage. In this type of tuner, switches connected to different voltage dividers can be used to change channels, or the voltages can be generated digitally using a D/A (digital-analog) converter. In order to receive a broad enough bandwidth, the tuner's passband extends a little into adjacent channels. The video IF circuits are provided with traps to eliminate the unwanted frequencies. A typical tuner frequency response for channel 2 is as shown in *Figure 9-2*. Tuners in color receivers always have an AFT (automatic fine tuning) circuit because the tuning must be exact; high enough to receive the color sidebands but not too high as to cause interference from the audio carrier. *Figure 9-3* shows a possible tuning arrangement for the VHF electronic tuner, using switches to select channels and pots for fine tuning. A varactor diode does not have enough tuning range to enable tuning through the full VHF band with a single

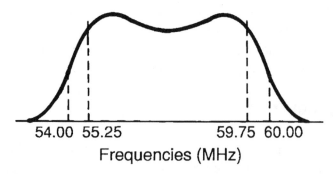

Figure 9-2. A typical tuner frequency response for channel 2.

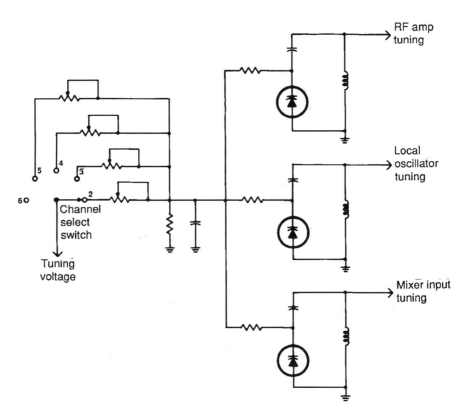

Figure 9-3. *A possible tuning arrangement for the VHF electronic tuner.*

coil. Switching diodes are used when tuning from low band to high band VHF that "short out" some of the coils' turns. *Figure 9-4* shows how the switching diodes connect to the coils in the previous diagram. When high band channels are selected, the switching diode is forward biased. A forward-biased diode has very low resistance, which effectively grounds the tap in the coil. The coil, now with fewer turns, has a lower inductance and resonates with the varicap at the high-band VHF frequencies. The UHF tuner, as shown in *Figure 9-5*, is made similarly, but the narrower tuning range (less than a 2:1 ratio between highest and lowest frequencies) makes it possible to tune the entire band without switching diodes. In the older, mechanically-tuned UHF tuners, resonant sections of the tuner were used as tuned cavities. Openings in the cavities coupled energy to each section and ganged variable capacitors varied the resonant frequency for each

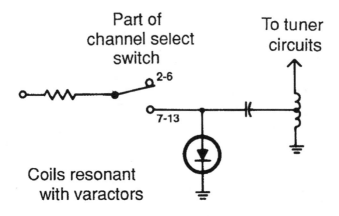

Figure 9-4. How the switching diodes connect to the coils in the previous diagram.

section. Most older UHF tuners were nothing more than three tuned sections (RF tuning, oscillator and mixer out), a transistor local oscillator and a hot-carrier diode for the mixer. When UHF was selected, the VHF tuner's mixer and, in some receivers, the RF amp became extra IF amps to boost the gain because there was no amplification in the UHF tuner. In tuners manufactured today, a single unit tunes both VHF and UHF. Most can also tune to the extra VHF cable channels. These tuners are compact and have no moving parts.

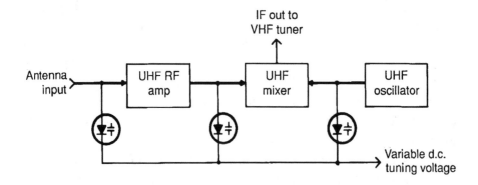

Figure 9-5. A UHF tuner.

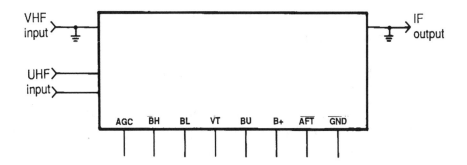

Figure 9-6. *The connections to the tuner will be identified in a way similar to that shown here.*

Tuner Terminals

The modern tuner usually consists of a small shielded box containing a single circuit board. All tuning and band switching is done with voltages input to the tuner's terminals. The connections to the tuner will be identified in a way similar to that shown in *Figure 9-6*. The VHF INPUT, UHF INPUT and IF OUTPUT are self explanatory. The other connections and voltages are:

(a) *AGC* - This is an analog (varying) voltage from the AGC circuits that controls the gain of the tuner's RF amps.
(b) *BH* - This is a digital voltage, meaning it is a voltage that will either be there, called "high," or not, called "low." The "high" can be around 5V to 15V, with 12V being common. When this pin is made to go high, high band VHF is selected.
(c) *BL* - Same as for BH, when this pin is made high, low band VHF is selected.
(d) *BU* - Digital voltage. Like BH and BL, when this pin is made high, UHF is selected.
(e) *VT* - This is an analog voltage, the tuning voltage. This voltage is developed by external circuits, possibly a microprocessor, that determines the frequency the tuner will receive.
(f) *B+* - This is the power supply voltage for the tuner, typically 12V.

(g) *AFT* - An analog voltage from an AFT detector or PLL circuit in the video IF stages keeps the tuner "locked in" to the correct channel frequency.

(h) *GND* - The ground connection for the tuner.

The AFT circuit may consist of a discriminator tuned to 45.75 MHz. If the video IF's picture carrier is at exactly this frequency, no control voltage will be developed. If the oscillator drifts or the fine-tuning is misadjusted, the picture IF will beat to a different frequency and the discriminator will develop an output voltage. This voltage will be fed to the local oscillator's varactor to either add to or subtract from its tuning voltage, depending on the direction of the drift. In many modern tuners, reference frequencies are generated and digital circuits used along with a PLL to keep the tuner locked in to precisely the correct frequency.

Video IF Amps

The video IF amps are basically small-signal tuned RF amps. In addition to amplifying the video and audio frequencies, these stages must have a wide bandwidth. In addition, they contain traps to eliminate the adjacent channel video and sound frequencies. A trap is also used at the sound frequency of 41.25 MHz to reduce the amplitude of the sound carrier to about 10% of that of the picture carrier. This is enough amplitude to demodulate the sound and helps prevent interference with the picture carrier. Ideally, the overall response of the amps should be as shown in *Figure 9-7*. In actual practice, this response is difficult to obtain. In addition to traps, the video IFs use stagger-tuned stages along with one or

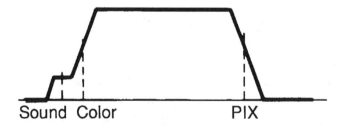

Figure 9-7. The overall response to the amps should be as shown here.

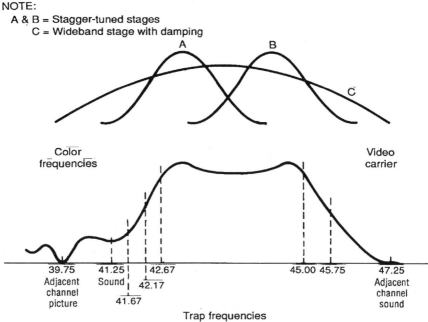

NOTE:
A & B = Stagger-tuned stages
C = Wideband stage with damping

NOTE:
(*top*) How adequate video IF response is obtained.
(*bottom*) The result - typical video IF response.

Figure 9-8. *Overcoupling may be used in one or more stages to widen the bandwidth.*

more wide band, heavily damped stages. In some, overcoupling may be used in one or more stages to widen the bandwidth, as shown in the upper drawing of *Figure 9-8*. The video IF stages would have a typical overall response similar to that shown in the lower drawing of the figure. All frequencies shown are, of course, in MHz. The dip, or "hump" in the response curve should not fall below 90% of the peak amplitude. The 41.25 MHz point should be at 10% amplitude. A typical transistor video IF amp stage is as shown in *Figure 9-9*. Many modern receivers have the video IF stages within integrated circuits, but will still have external tuned circuits and traps.

Video Amplifiers

Video amplifiers are basically high quality wideband amplifiers. Video amplifiers must have approximately a 4 MHz bandwidth. They must also

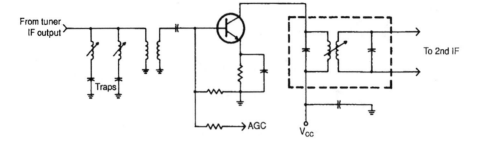

Figure 9-9. A typical transistor video IF amp stage.

have the proper phase output, and must not introduce phase distortion or "lag." If the signal passes through an RC network, lower video frequencies will exhibit more phase shift than the higher frequencies because of the greater reactance. This amounts to a small difference in time between input and output. Since it takes only 63.5 seconds to scan one horizontal line, a small amount of lag can cause a problem, causing objects in the picture to be displaced to the right. This can produce a "smear" in the picture. Where possible, direct-coupled video amps are used to eliminate the reactive phase lag. To obtain the proper bandwidth, several things may be done. The video amps may use lower values of load resistances to produce an overall wider bandwidth. Since gain is proportional to the load resistance and the gain-bandwidth product of an amplifier is a constant, lowering the gain serves to widen the frequency response of the amplifier. The drawing in *Figure 9-10* illustrates this.

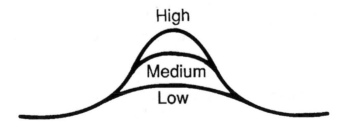

Figure 9-10. Lowering the gain serves to widen the frequency response of the amplifier.

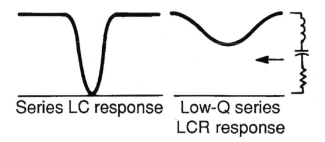

Series LC response   Low-Q series
LCR response

Figure 9-11. A shunt peaking circuit.

Obtaining a "Flat" Response

The response will still taper off on the upper and lower ends of the band. If we had a load resistance that had a higher impedance on the upper and lower ends and lower impedance in the middle of the curve, we could get a gain compensating curve to flatten the response. This can be accomplished with a low Q series resonant circuit, or series LCR. This circuit, shown in *Figure 9-11*, is called shunt peaking because the circuit is used in parallel with the output of one of the video amps. Series peaking circuits are also used. This type improves only the high frequency response. Series peaking uses an inductor to resonate with the input capacitance of a video amp stage. If the resonant frequency is chosen above the video

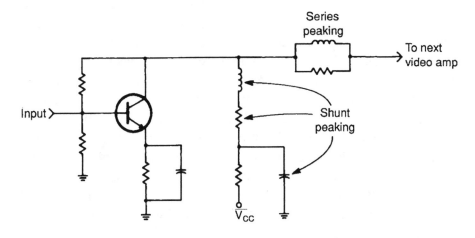

Figure 9-12. A video amplifier employing both series and shunt peaking.

passband, a rising response will be observed at the upper frequencies as this frequency is approached. In order to prevent a sharp rise from occurring, swamping resistors are often used across the coil. The peaking circuits in video amps are not normally adjustable, but some receivers have a variable resistor in the peaking circuit called a sharpness control. *Figure 9-12* shows an example of a video amplifier employing both series and shunt peaking.

Video Amplifier Design

The emitter follower is often used for video amplifiers and can be used without frequency compensation because of its low output impedance. It cannot provide voltage gain, however, and is sometimes followed by a common base amp to provide this gain. This works well because the common-base amp's low input impedance can closely match the output impedance of the emitter follower. A common-emitter amp direct coupled to a common base amp also works well. With this arrangement the two devices are essentially in series. This is often referred to as a "cascoded" video amplifier.

The DC Restorer

If capacitive coupling is used in any of the video amplifiers, the DC component of the video signal will be lost. If this occurs, a change in peak value of the video signals will cause the value at both extremes of the signal to change. For example, if a brightly colored object appears in the video, the negative voltages (representing darker objects) would become more negative to compensate and maintain an average of 0V. The gray tones might then become black, and the darker shades might be lost altogether. To compensate, the DC component of the signal can be reinserted by a DC restorer or clamp circuit. A diode clamp circuit is often used that clamps the video to 0V at the peak sync level. Of course, if direct coupling is used throughout, dc restoration is not needed. Many smaller monochrome receivers do without dc restoration by adding a fixed amount of dc to the video. The peak levels will still vary but modern CRTs have better contrast ratios that can still produce acceptable pictures this way.

Sync Circuits

The sync circuits consist of the sync separator and sync amplifier. Some receivers may not use a sync amp if the level of sync from the sync separator is sufficient. The sync separator separates the sync from the video so that it can be used by the sweep circuits. *Figure 9-13* shows the basic operation of the sync separator. If the DC voltage at the blanking level is 2V, as shown, for most of the video the voltage will be less positive than 2V. Since the cathode of the diode is connected through a resistor to a positive 2V, it will be reverse biased during this time and will not conduct. The only time it can conduct is during the sync pulses, when the anode voltage exceeds 2V. We are assuming for this example that this is an ideal diode. In actual diodes the barrier voltage would have to be taken into account. This circuit is not used in practice because the video signal is subject to vary somewhat. The circuit can be modified, however, so that the bias voltage for the diode varies with the peak level of the video as shown in *Figure 9-14*. For this circuit, as the diode conducts the capacitor develops a charge equal to the peak sync level. The time constant is such that the capacitor discharges only a small amount, not enough to

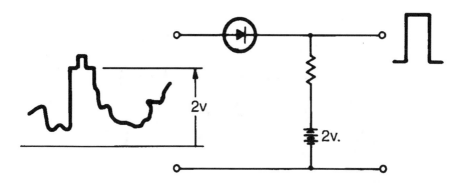

Figure 9-13. The basic operation of the sync separator.

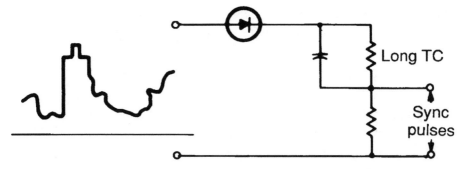

Figure 9-14. *The circuit can be modified so that the bias voltage for the diode varies with the peak level of the video.*

bring the voltage down to the blanking level, after one complete horizontal line. At this time another sync pulse occurs which now exceeds the capacitor's charge by a small amount so the diode does conduct during this time and produce an output pulse. For longer term variations the capacitor's charge can adjust to the signal level.

An Amplifying Device as Sync Separator

Although the diode sync separator works, it does not produce a square pulse. It is much more common to use an amplifying device, such as a

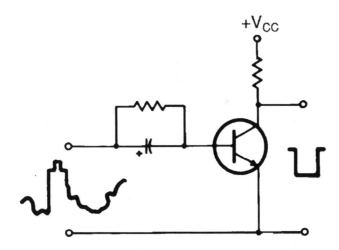

Figure 9-15. *A simplified diagram of a transistor sync separator.*

transistor, as the sync separator. It can be biased so that it is driven between saturation and cutoff by the sync, thus squaring off the pulses. It also provides a larger amplitude for the sync and, in many cases, makes a sync amp unnecessary. *Figure 9-15* is a simplified diagram of a transistor sync separator. Most color receiver sync circuits use a sync amplifier following the sync separator.

The Noise Gate

Some receivers employ a noise gate to prevent impulse noise, such as static or lightning, from interfering with the sync. If the noise were to be sufficient in amplitude to exceed the sync level, it would cause the picture to roll or tear. The diagram of *Figure 9-16* is an example of a noise gate. In this diagram, Q1 is the sync separator and Q2 the noise gate. Notice that two opposite-phase video signals are used for this particular circuit.

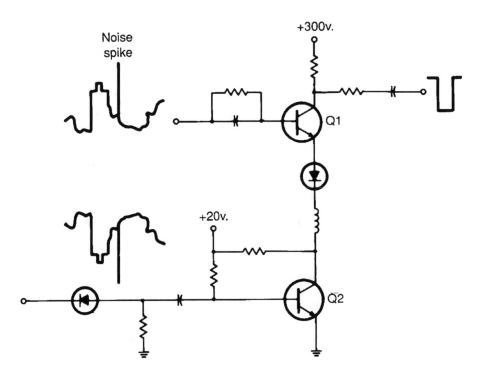

Figure 9-16. An example of a noise gate.

The bottom transistor is biased such that it is normally saturated and thus connects the emitter of the sync separator to ground. A signal that is sufficiently negative, like the noise spike shown, can override the positive voltage causing Q2 to go into cutoff during that time. This effectively disconnects the emitter of the sync separator from ground and does not allow it to work during this time.

AGC Circuits

It is apparent that differences in reference levels for the video signal cannot be tolerated to any great extent. Therefore, the carrier level at the video detector must remain fairly constant, regardless of the received signal strength. AGC circuits must develop two different AGC voltages. One is for the tuner, called the RF AGC voltage, and the other is the IF AGC voltage for the video IF amps. Since the signal levels at the tuner input are very small, much less variation in the AGC voltage is needed there. Resistors may be used to lower the IF AGC voltage to the correct value for the RF AGC. Variable resistors may be used for these to allow

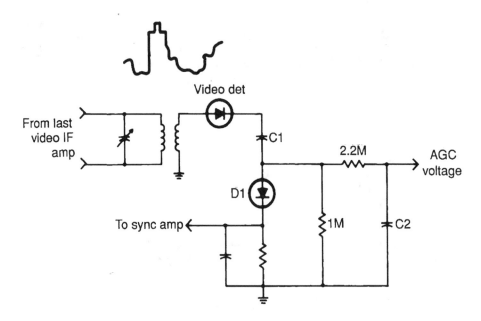

Figure 9-17. A simplified diagram of a peak AGC circuit.

the AGC voltages to be set to the optimum point. The AGC detector works similarly to the sync separator and in some receivers may be integral with it. *Figure 9-17* is a simplified diagram of a peak AGC circuit. The only change from the operation of the sync separator is that another RC circuit is added with a much longer time constant. C2 is capable of holding a dc charge equal to the peak sync for several horizontal lines. This produces a smooth output voltage. This voltage will follow any longer term changes in peak amplitude. In some cases this control voltage may not be large enough to control the gain of the IF amps, so an AGC amplifier may be used. The circuit of *Figure 9-18* is an example. Some earlier receivers used a local/distance switch to select low or high gain for the AGC amp. This helped prevent overloading of very strong signals or reducing the gain too much for very weak ones. The peak AGC uses a long time constant and thus cannot follow signals that change quickly in signal strength. This can cause the picture to "flutter" when the signal reflects from cars, trains or other moving objects. Another problem it has is that it is responsive to any peak signal. This can cause noise spikes to "set" the AGC level and weaken the actual video when noise occurs.

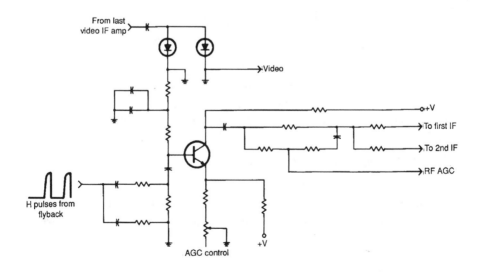

Figure 9-18. The control voltage may not be large enough to control the gain of the IF amps, so an AGC amplifier may be used.

Keyed AGC

Most receivers today use a keyed AGC system. The keyed AGC is an improvement because it is made to be responsive to the signal only during the sync. It is then not responsive to noise during horizontal scanning time. It can also use a much shorter time constant during the sync time allowing the AGC voltage to adjust quickly. Usually, pulses from the flyback transformer are used as the supply voltage for the AGC stage so that it is only operative during this time. *Figure 9-19* is a circuit example of a keyed AGC stage. In most receivers made today, the AGC circuitry is contained within an integrated circuit along with the sync circuits and possibly, many other circuits.

The Color and Brightness Circuits

The color and brightness will be covered in a future article segment in a future issue.

Color Television Receiver Circuits: Part 3
By Lamarr Ritchie

This is the third part of a four-part article that describes all of the circuits in a modern TV receiver. The fourth and final part, deflection circuits and high voltage, will be published in the July 1996 issue of *ES&T*.

Chroma Bandpass Amps

The chroma bandpass amps remove the color portion of the video signal and amplify it in addition to the color burst. The key points concerning chroma bandpass amps are:

(a) They are tuned amplifiers with a bandwidth of at least 1 MHz.
(b) The last bandpass amp before the demodulators may be keyed off during retrace to prevent the burst from reaching the demods.

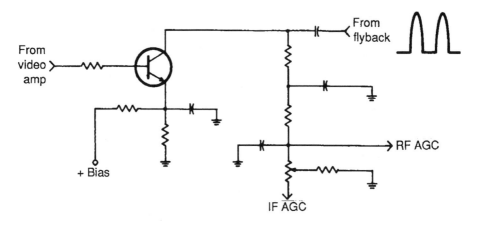

Figure 9-19. A circuit example of a keyed AGC stage.

(c) The level control for the color will be found in one of these stages, or at their output for the user to adjust the saturation, or amount of color.

(d) Two voltages associated with the bandpass amps are the color killer bias and ACC (automatic color control) bias.

(e) Peaking is used at about 4.08 MHz to compensate for the sloping part of the video IF curve that contains the color, as shown in *Figure 9-20. Figure 9-21* illustrates a typical chroma bandpass amp.

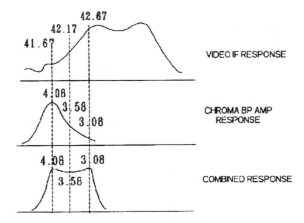

Figure 9-20. Peaking is used to compensate for the sloping part of the video IF curve that contains color.

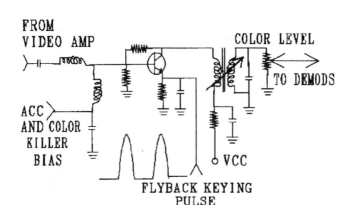

Figure 9-21. A typical chroma bandpass amp.

Burst Amplifier

The burst amp removes and amplifies the color burst used to lock in the color oscillator. To enable it to amplify only the color burst, it must be a keyed amplifier, keyed on during the blanking interval when the burst occurs. It may be keyed using pulses from the flyback transformer, or delayed pulses from the sync separator. Keying from the sync has the advantage of making the timing independent from the horizontal hold control or AFC circuits which could change the phase of the amplifiers output and therefore, the hue of the picture. In the example circuit of *Figure 9-22*, the keying pulses provide the base bias voltage to the transistor allowing it to amplify the signal supplied from the chroma bandpass amp during this time. Since it is the color burst that occurs during this time, this is the only signal that will appear at the output of this stage.

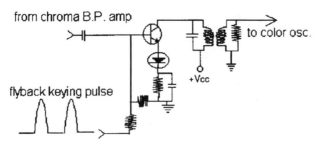

Figure 9-22. The keying pulses provide the base bias voltage to the transistor allowing it to amplify the signal supplied from the chroma bandpass amp.

Color Oscillator

The color oscillator is a crystal oscillator at 3.579545 MHz. This oscillator is used to reinsert the color subcarrier that was suppressed by the color camera circuitry. The CW from the oscillator will be used as a reference in the demodulators to recover the color video signal. A crystal oscillator alone is not stable enough to mimic the frequency and phase of the original subcarrier, so it must be used in conjunction with the color sync to provide a phase-locked reference signal. Associated with the color oscillator, or burst amp feeding the oscillator, is the hue control for the color. Varying the phase of the oscillator CW will vary the phase of the demodulated R-Y and B-Y signals, thereby varying the hue or tint of the color. It is well to note here that any internal adjustments of the bandpass coils and transformers associated with either the burst amp or oscillator will vary the phase, and hue, of the color signal.

Chroma Demodulators

The chroma demods take the color signal input from the chroma bandpass amps along with the CW signal from the color oscillator and recover the original color difference signals. TV receivers have either two or three demods. Most receivers have R-Y and B-Y demods to recover these two original signals. In most older receivers, a matrix circuit mixed these two in proper amounts to produce the G-Y signal. Obtaining the G-Y signal in this way requires fewer parts and is a little cheaper and easier to accomplish where integrated circuits are not used. Most modern receivers have a G-Y demodulator instead of the matrix circuit. Some receivers used X-Z demodulators. These altered the phase angles to make it easier to recover the G-Y signal. In addition to demodulating the color signal, the demods must also restore the proper amplitude to the color difference signals. These were altered during the modulation process. Since blue colors tend to be darker, if the correct amplitude of the color signal for a saturated blue were used, the negative half-cycle of the color signal would extend well into the sync region. To prevent this from happening the B-Y signal is reduced in proper amplitude. Likewise, hues containing red are reduced in amplitude, but not as much as the blue. Since the eye is more sensitive to hues containing green, and green is the major component of white light and thus the luminance signal, the G-Y signal is increased in proper amplitude. The actual changes in proper amplitude for the three signals are: B-Y 0.49, R-Y 0.877, and G-Y 1.423. The receiver's demods must have their gain altered to compensate, so their relative gain must be: B-Y 1/0.49 = 2.03, R-Y 1/0.877 = 1.14, and G-Y 1/1.423 = 0.70. Basically, a color demod is a differential or additive type amplifier with the color signal at one input and the 3.58 MHz oscillator CW at the other. As the phase between the two vary, the output amplitude will vary. Although we will be discussing some of the discrete circuits, as we have been doing, it must be pointed out again that modern receivers have these circuits within integrated circuits. The "color IC" for a TV may contain all of the color circuits from bandpass amps through demodulators. An understanding of the circuit operation is still necessary to assist in recognizing color problems and diagnosing problems in discrete components in the color circuits. *Figure 9-23* is an example circuit of a B-Y demod. The two signals are fed to the transistor as shown. When the phase of the chroma is at 0 degrees, it is in phase with the oscillator CW and minimum output current occurs because there is little difference in voltage between the transistor base and emitter. This will produce maximum collector voltage at this point. If the chroma signal were to be at 180 degrees, roughly corresponding to the color yellow, then maximum collector current will flow producing minimum collector voltage. The diagram shows the relative outputs that would occur for the colors yellow, white and blue. The

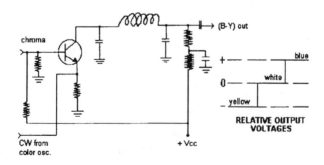

Figure 9-23. An example circuit of a B-Y demod.

output circuit of the demods is a low-pass filter with a cutoff frequency of about 0.5 MHz to 1.5 MHz, depending on the receiver. This filter eliminates the 3.58 MHz signals and produces only a varying signal voltage that corresponds to the average of the total difference in phase of the two input signals. The smaller the phase angle between the two signals, the greater the output voltage. Keep in mind that the encoded color signal is input to the demods. This is not video. The output of the demods is a video signal. For input to the grid of the CRT, the above circuit has the proper phase, since positive video must be used there. However, for cathode drive the signal must be of negative phase. That is, we will need -(B-Y) instead of B-Y. This is no problem because the following stage can invert the phase of the signal.

Transistor Chroma Demods

The only difference between this demod and the R-Y demod is that the CW signal is shifted in phase by 90 degrees before being input to the demod. Then, if the color signal phase is near 90 degrees it will be in phase with the oscillator CW, and maximum collector voltage will be produced. *Figure 9-24* is an example diagram of transistorized R-Y and B-Y demodulators. A circuit known as a balanced demodulator circuit has been used in some receivers and can be implemented using diodes instead of transistors. Although a separate demodulator can be used for the G-Y signal, it has been noted that this signal can be derived by mixing the R-Y and B-Y signals in the proper proportion. This is possible be-

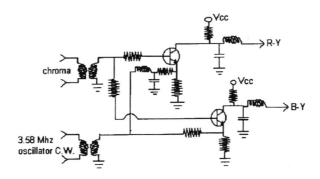

Figure 9-24. *An example of transistorized R-Y and B-Y demodulators.*

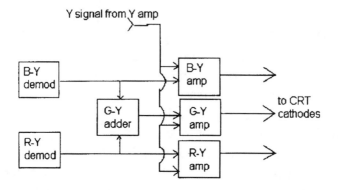

Figure 9-25. *A method of obtaining the G-Y signal.*

cause all three color signals were used in a mixture to produce the two color difference signals. *Figure 9-25* shows a method of obtaining the G-Y signal. This diagram also shows a method of mixing the luminance (Y) component back in to the color difference signals. This is the most often used method. In this method, one of the two signals (Y and color differ-

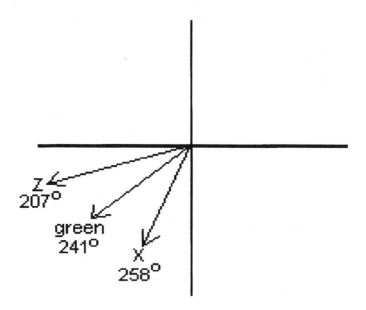

Figure 9-26. *The phase angle of the two demodulators, compared to the G-Y signal.*

ence) may be fed to the base of the amplifier transistor and the other to the emitter. The output would then contain R, G, or B with the proper amount of luminance. Matrixing, or mixing of the signals can be accomplished at the picture tube also, using both the grids and cathodes of the CRT. For instance, the separated color difference signals (at a negative phase) can be tied to the individual cathodes. The luminance signal (at a positive phase) from a Y output amp can be connected to the three control grids, the grids all being connected together. To help in the mixing process to obtain the G-Y signal, phase angles other than 0 degrees and 90 degrees can be demodulated in an X-Z demodulator. The phase angle of the two demodulators, compared to the G-Y signal, is shown in *Figure 9-26*. With the vectors positioned as shown, equal amounts of the X and Z signals produce the G-Y signal. *Figure 9-27* is a simplified diagram illustrating how this is done. The altered phase angles allow the G-Y signal to be developed across the common emitter resistor. The Z signal contains -(B-Y) and G-Y. The X axis contains -(R-Y) and G-Y. The signal common to both will be present at the emitters. This signal will then cancel the G-Y signal in the two demods. (The emitter signal is opposite in phase to that of the base.) The phase shift in the transistors changes the -(B-Y) and -(R-Y) to B-Y and R-Y. The G-Y at the emitters is already at

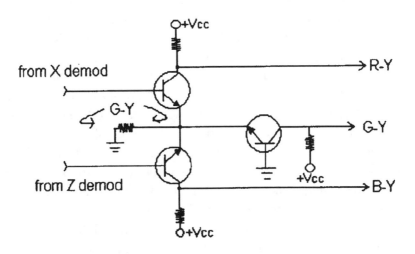

Figure 9-27. *A simplified diagram illustrating how equal amounts of the X and Z signals produce the G-Y signal.*

the proper phase so the common base G-Y amp does not invert its phase. There is an added advantage to this arrangement in that, sharing common signals at the emitters, changes in one are more likely to affect all three and the color will remain more accurate. In color demods, no matter which type, a change in phase of the C signal varies the output of the demods in different proportions producing different mixes of the colors and therefore, different hues. A change in amplitude of the C signal, however, changes the output of each in the same proportion, changing the color saturation but not the hue. The Y signal, in some cases, is mixed back in with the color difference signals at the demods. The video amps following the demods are called either the color difference amps (R-Y, G-Y or B-Y amp) or color video amps (R, G or B amp) depending on whether their output has the luminance mixed back in or not.

Blanker Stage

Although the video signal was designed with blanking pulses to turn off the CRT during retrace, these pulses are not really sufficient to do this. For them to work it would require very accurate adjustment of internal and consumer controls. The blanking pulses are only 7.5% "blacker" than the black level of the video. If the brightness were adjusted so that blacks were not really black, but a gray tone, then the blanking might be above the black level. Retrace lines would then be visible on the CRT screen. To keep this from happening, TV receivers usually use high amplitude pulses from the deflection circuits as blanking for the CRT. Vertical pulses may be fed to video amp stages or the CRT. High amplitude negative pulses may be fed to the screen grids of the CRT to cancel the positive screen voltage during vertical retrace. To provide blanking of horizontal retrace, a blanker stage is often used. An example of a blanker stage is shown in *Figure 9-28*. In this example, the positive flyback pulses are inverted in phase by the blanker's common emitter configuration and applied to the emitters of the color difference amps. Being input to the emitter, they will not be inverted in phase by these amps, and so will still be negative at the CRT grids, putting the CRT in cutoff at this time. A blanker stage is sometimes used for vertical blanking also, as is a blanker that is fed pulses

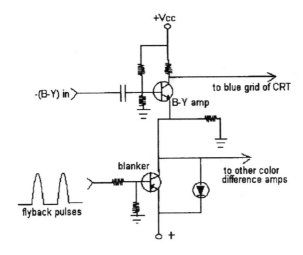

Figure 9-28. A blanker stage.

from both deflection circuits through a resistive matrix circuit. In some receivers the amplitude from the blanker is adjustable and is adjusted to eliminate retrace lines in the picture.

Color Killer and ACC (Automatic Color Control)

The color killer and ACC circuits are related as shown by *Figure 9-29*. The purpose of the color killer is to disable the color bandpass amps during a monochrome broadcast. If the color circuits were not disabled, random noise might cause colored snow or "confetti" on the screen that would be objectionable. The color killer senses the presence of the color

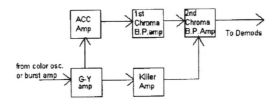

Figure 9-29. The relation between the color killer and ACC circuits.

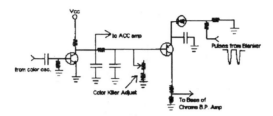

Figure 9-30. *The color killer senses the presence of the color burst, either out of the burst amp or the color oscillator.*

burst, either out of the burst amp or the color oscillator. *Figure 9-30* is an example. For this circuit, the killer amp supplies the bias for the chroma bandpass amp. If the color burst is present, the output of the color oscillator increases somewhat and biases on the killer detector, which sends bias to the bandpass amp, allowing it to operate. If the burst is not present, the killer amp will not be biased on and will not develop the required bias for the bandpass amp. It will thus be disabled. The ACC circuit works similarly, the only difference being the amount and polarity of bias voltage developed. This circuit (shown in *Figure 9-31*) connects to the previous diagram labeled "to ACC amp." At that point, an increase in the color signal will cause a decrease in voltage because of the increased conduction of the transistor. This "less positive" voltage will cause the PNP ACC amp to conduct more heavily. This, in turn, causes the emitter of the NPN bandpass amp to become more positive, thereby lowering its gain. This circuit, then, is actually an automatic gain control for the color signal.

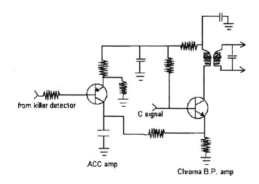

Figure 9-31. *An ACC circuit.*

Automatic Tint Control (ATC)

Color circuits such as these may be known by different names in different brands of receivers. This circuit acts to keep the hue of the color constant for different color broadcasts. Actually, it works to keep the hue of the flesh tones constant, regardless of the other colors, since it is the most important recognizable color. These circuits are not as important today as they were in earlier receivers because the broadcast equipment and standards are much better now. Two general methods are used for the ATC: (a) Emphasize the red hues by increasing the gain of the R amp or decreasing the gain of the B amp; (b) Shift the phase of the B-Y demod closer to the phase of the R-Y. Refer to the circuit example of *Figure 9-32*. When the ATC switch is turned on, Q1 conducts causing Q2 to saturate. This effectively ties the end of C1 to ground causing a phase shift in the B-Y demod. Also, the end of R1 is now effectively grounded through the saturated Q2 and being in parallel with the emitter resistance of the red amp, raises its gain. Not to be confused with these circuits, some receivers have factory preset adjustments for controls like the brightness, contrast, color and hue. A switch on the TV (it is known by various names depending on the manufacturer) will throw in all factory preset controls in lieu of the consumer controls. This would prevent the customer from having to make any adjustments to the television if these settings are acceptable. These controls are sometimes located on the chassis and not easily accessed for adjustment. Some, however, are placed where

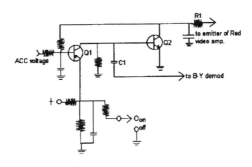

Figure 9-32. An ATC circuit.

they can be easily changed. For example, they might be ganged to the main control such that a small screwdriver can be inserted through a hollow shaft of the main control to adjust the preset control.

Automatic Brightness Level (ABL)

This type of circuit is often used in color receivers to improve the quality of the picture. The ABL generally does two things. It maintains the correct brightness with changes in AC line voltage, and maintains the correct levels for changes in CRT anode current. *Figure 9-33* is a circuit example. Note the operation of the brightness control here. The control is varying the bias of the video amp. As the wiper moves up, the base of the amp is made more positive (less negative) and the transistor conducts less, making the emitter go more positive. This at the cathode of the CRT makes the picture darker. Moving the wiper down would have the opposite effect and make the picture lighter. One of the supplies for this control is an unregulated supply. If the line voltage rises this would tend to increase the high voltage and video levels and make the picture brighter, but a rise in positive voltage at the base makes the video amp conduct less and raises its emitter voltage. This has the effect of darkening the picture and counteracts the increase. The regulated supply is mixed with the unregulated voltage to -tone down" this effect for the correct amount. The brightness limiter transistor regulates against changes in beam current

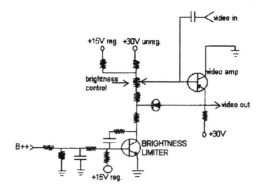

Figure 9-33. An ABL circuit.

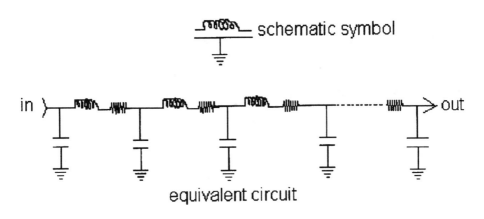

Figure 9-34. *A video delay line.*

using the B++ voltage. This voltage, mixed with the +15V regulated supply normally keeps the transistor saturated. If there is excessive brightness, causing too much beam current, the B++ voltage will fall because of the excessive current drain. This will cause the transistor to come out of saturation and its collector voltage will rise causing the base voltage of the video amp to become more positive. As seen before, this will darken the picture and tends to limit the maximum brightness to the point where excessive beam current would cause the high voltage to fall.

Video Delay Line

The video delay line is in the luminance signal path to match the delay in the color circuits. The delay is about 0.8 microseconds. The delay line is a coil wound over an insulating tube and foil strip. The effect is like having a capacitor in parallel, to ground, for each turn of the coil as shown in the equivalent circuit of *Figure 9-34*. The resistance of the wire is also a part of the equivalent circuit. The delay line must be terminated with its characteristic impedance, typically 600Ω to 700Ω. Its DC resistance is usually about 100Ω to 150Ω.

Deflection and High Voltage

The final installment of this series, scheduled for publication in the July issue, will cover deflection and high voltage.

Color Television Receiver Circuits: Part 4

By Lamarr Ritchie

Three earlier articles in this series have described most of the circuits in a typical color TV set. This fourth article wraps things up with a discussion of deflection circuits and the high-voltage section.

Deflection Circuits

The deflection circuits, both horizontal and vertical, develop the scanning currents for the electron beam. The end device for both vertical and horizontal deflection circuits is the deflection yoke, which is situated around the neck of the picture tube. The deflection currents are produced by oscillators. The two most popular types of oscillators used as deflection oscillators are the blocking oscillator and the multivibrator. In transistor type deflection circuits, an emitter follower driver amp is often used to prevent loading of the sawtooth forming circuitry of the vertical oscillator.

Vertical Deflection

The vertical deflection circuit example shown in *Figure 9-35* illustrates two adjustments commonly found in the circuits. The vertical height control adjusts for the correct amount of vertical sweep. This example is adjusting the amplitude of the oscillator output voltage to do this. The vertical linearity control adjusts the linearity of the sawtooth, making sure the top or bottom of the raster is not stretched or compressed. The vertical output stage may be followed by a vertical output transformer to

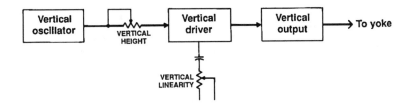

Figure 9-35. *In this circuitry, the vertical height control adjusts for the correct amplitude of vertical sweep, and the vertical linearity control adjusts the linearity of the sweep sawtooth, insuring that the top and bottom of the raster are not stretched or compressed.*

match the impedance to the yoke. Most transistor vertical outputs do not employ a transformer, instead they use an amplifier arrangement such as the complementary symmetry amp that has a low output impedance. To separate vertical and horizontal sync, integrators and differentiators are used. Their operations are illustrated in *Figure 9-36*. The vertical integrator may have more RC sections than the one shown, and may be in packaged component form, having three leads; input, output and ground. The output of the vertical integrator is sufficient to lock in the vertical oscillator because, being essentially a low pass filter, it does not produce an output for fast-acting or high frequency noise.

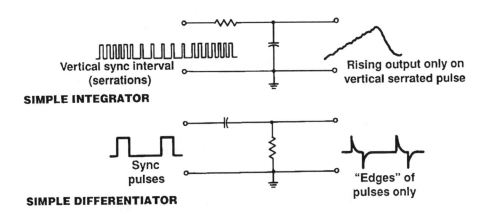

Figure 9-36. *Integrators and differentiators are used to separate vertical and horizontal sweep.*

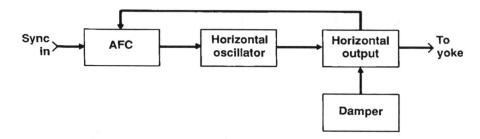

Figure 9-37. The horizontal sweep section of a modern TV is generally based on a scheme such as this.

The Horizontal Oscillator

The horizontal oscillator, however, needs a better way to control its frequency since high frequency noise will be present at the differentiator's output. It is essentially a high-pass filter. The horizontal circuit uses a phase-locked loop called the horizontal AFC circuit to control the oscillator's frequency. *Figure 9-37* shows the basic configuration of the horizontal sweep section. The horizontal oscillator is locked in by a DC control voltage developed by the AFC. The control voltage developed by the AFC's phase lock circuit can be a smoother, slower moving voltage with low pass filtering to remove the effects of high frequency noise spikes.

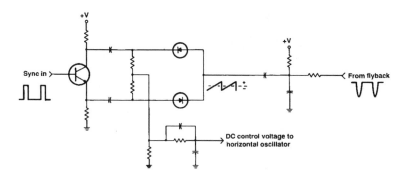

Figure 9-38. The AFC circuit keeps the picture synchronized by comparing the arrival times of the horizontal sync pulses with the time of occurrence of the 0V point in the horizontal sawtooth waveform and automatically adjusting them so that they occur at the same time.

Automatic Frequency Control (AFC)

Figure 9-38 is an example of an AFC circuit. The input transistor in this circuit acts as a phase inverter to provide two equal pulse inputs of the correct phase to make both diodes conduct. The phase inversion at the collector provides the cathode of the top diode with negative pulses. The signal at the emitter is not phase shifted, so the anode of the bottom diode is provided with positive pulses. If the voltage at the other end where the diodes join is at 0V, both diodes will conduct equally. The divider resistors for the control voltage will then have equal voltage drops, and being opposite in phase, will cancel and the control voltage will be 0V. This situation occurs if the sawtooth derived from the flyback is at the 0V point, as shown by the slashed line, at the same time as the input sync pulses occur. If however, the sawtooth arrives earlier or later than the sync pulses, it will not be at the 0V point. If the frequency of the oscillator is too high, the pulses will arrive too early and will be on the positive part of the slope when the sync pulses arrive. This will cause more forward bias for the top diode, and less for the bottom diode. The two resistor currents will then be unequal and will not cancel. A control voltage will thus be generated that, in this case, will be negative. If the frequency of the oscillator is too low, the reverse of the above situation will occur and a positive control voltage will develop. The control voltage is connected to the oscillator to cause its frequency to vary in the proper direction until the control voltage falls to zero. The oscillator stays phase locked to the sync signal.

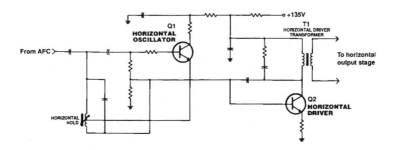

Figure 9-39. *The horizontal oscillator generates the sawtooth waveform.*

Multivibrator Characteristics

Multivibrators are very sensitive to voltage changes. The control voltage could possibly vary the frequency of the multivibrator too much to be used as is, so a tank circuit can be added to the multivibrator to stabilize the circuit. Changes in voltage do not affect the LC oscillator to a great degree. In the example circuit of *Figure 9-39*, Q1, the horizontal oscillator, and Q2, the horizontal driver, actually form a multivibrator. There is a feedback path from the collector of Q2 to the base of Q1. This is used in combination with the Hartley oscillator arrangement to provide the proper stability. In this circuit the horizontal hold control adjusts the core of the coil to change the free-running frequency of the oscillator. Transistor Q2 will in most cases be an intermediate power transistor, since this stage must drive the relatively high-powered horizontal output stage. An interstage transformer, called the horizontal driver transformer, is usually used to match the driver output impedance to the horizontal output's input impedance. The input pulses to the horizontal output transistor will not be a sawtooth wave, they will be rather narrow rectangular pulses. Capacitors at the collector will give a small sawtooth component to the output voltage, but the signal will still consist mainly of rectangular pulses. The time constant of the flyback transformer and yoke, being a relatively

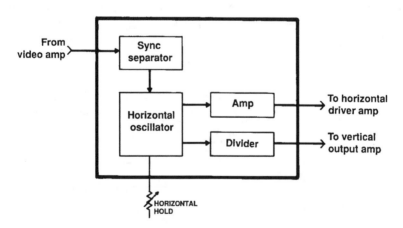

Figure 9-40. Most modern TV receivers have all of the sync and deflection circuitry, except for the power stages, in a single integrated circuit package.

high Q, will generate a sawtooth current when a rectangular pulse is applied. The small amount of sawtooth voltage will provide a sawtooth current for the small amount of circuit resistance. Most modern TV receivers now have the entire sync and deflection circuits, except for the power stages, within a single integrated circuit. (*Figure 9-40.*) In these circuits, there is no vertical integrator or oscillator and therefore no vertical hold control. The vertical frequency is derived by digital circuits that divide the horizontal oscillator's phase-locked signal by a factor of 262.5. There are some modern hybrid ICs that also contain the vertical output stages and horizontal driver stage, but they all still have a discrete horizontal output stage using a transistor.

The Horizontal Output Transistor (HOT)

The horizontal output transistor is a high voltage, high current switching transistor. The HOT will usually be rated at 1500V to 2500V at 5A to 15A, and are almost invariably NPN transistors. Take special note of *Figure 9-41*. It is important to be aware that there are special horizontal output transistors such as this that contain an integral damper diode. Some also contain one or two circuit resistors within the package. If you try to test one of these HOTs as you would an ordinary transistor and are not aware that there are other devices in the package, the transistor might appear defective. If in doubt you must obtain the information from a substitution guide or the service manual. The horizontal output stage is generally a high power, high frequency amp. The sawtooth wave has many harmonics starting with the fundamental frequency of 15.75 KHz. The horizontal output stage is biased class C for best efficiency.

The Damper Diode

Because of the high inductance and inter-winding capacitance of the yoke and flyback transformer, the circuit must be damped out during retrace to prevent ringing. The damper diode does this. The damper diode is placed in parallel with the high inductance flyback, in most cases directly across the horizontal output transistor. As shown previously, the damper may be within the package of the horizontal output transistor. During flyback time, the reverse voltage (sometimes called counter EMF, or CEMF) drives the collector of the horizontal output negative, so it stops conducting and will be of relatively high impedance. This would allow the L and C of the circuit to be shock excited and begin to ring. The high CEMF would also destroy the horizontal output transistor. The damper diode is connected in such a way that it can conduct during this time and present a low impedance across the windings to prevent this from occurring. Although the high CEMF will not appear across the primary winding of the flyback because of the damper, it will appear across the secondary winding and is used for the high voltage. The Q of the circuit is high enough that it takes some time for the CEMF to die out and the voltage to go positive again at the primary. Hence, the damper is conducting and providing deflection current for the first part of the sweep, as well as during the retrace. In tube-type receivers the damper diode supplied up to 30% or 40% of the horizontal sweep, but solid state damper diodes do not supply as much of

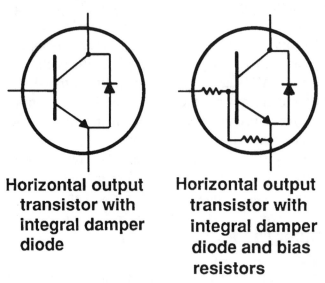

Horizontal output transistor with integral damper diode   **Horizontal output transistor with integral damper diode and bias resistors**

Figure 9-41. Many horizontal output transistors (HOTs) contain a damper diode, and sometimes bias resistors, within the same package. If you try to test one of these devices as if it were a standard transistor you might get a reading that would lead you to the conclusion that it's defective even though it is in perfect shape.

the horizontal sweep because they are lower impedance devices. Damping is not critical in the vertical section because vertical retrace is much slower, slower even than horizontal trace. A small amount of damping may be provided by resistors across the yoke windings.

High Voltage Section

The high voltage section, for simplicity, can be considered to be integral with the horizontal sweep section. The high voltage pulses are produced as a by-product of the quick horizontal retrace time. To step this voltage up even further, a step up winding is used in the flyback transformer. The high voltage rectifier then changes the AC pulses to DC. Because of the capacitance of the CRT, no filter capacitor is needed. Extreme caution must be used when working around the high voltage section. The voltage in this area ranges from 10KV or less for a small monochrome receiver, to 30KV or more for a color receiver. Voltage this high can ionize the air and thus form a current path, or arc for one to two inches. Moreover, the capacitance of the CRT can store this charge for more than 24 hours after the TV is turned off. Many receivers have bleeder resistors for the high voltage, but you should always discharge the anode lead to ground before touching anything in the high voltage section. Some old receivers used a high voltage section independent of the horizontal, called a "RF high voltage supply," but these are rarely seen today.

The Flyback Transformer

The flyback transformer is the central component in the high voltage/horizontal section, and is second only to the horizontal output transistor in being the most prone to failure. It usually has many functions and a failure in the horizontal circuits may disable every other circuit, causing the receiver to be completely dead. In addition to sweep and high voltage, the flyback transformer supplies keying pulses for various circuits, blanking pulses, horizontal AFC pulses, etc. It may also provide the boosted B+ (B++) to be used as screen voltage for the CRT screen voltages and for other circuits. As mentioned, the flyback supplies keying pulses, blanking pulses, horizontal AFC pulses, etc. It provides the focus voltage (first anode voltage) for the CRT. This voltage will usually be in the range of 2KV to 4KV. *Figure 9-42* is a simplified schematic example of an HV/ horizontal output section.

Regulation

All color receivers require the high voltage to be well regulated. Some older models regulated the high voltage directly, but most receivers today regulate it indirectly by regulating the DC supply voltage for the horizontal output stage, which will normally be in the range of 80V to 120V.

Supply Voltages Derived From the Flyback

Most TVs obtain their DC power supply voltages by using additional secondary windings on the flyback transformer. This eliminates an additional power transformer that would add weight and cost. The voltages from the flyback transformer are easier to filter because of the higher frequency of the horizontal pulses. A disadvantage of deriving supply voltages from the flyback—in addition to added complexity—is the possibility of costly damage to the flyback, horizontal output, and other horizontal sweep components if any other circuit fails, drawing too much current. Moreover, a failure in the horizontal circuit causing the horizontal output voltage to increase, or the frequency to change, could cause the voltage to all other circuits to increase and cause major damage.

Hold-Down and Shut-Down Circuits

Most receivers have additional protection circuits. These may be called "hold down," or "shut down" circuits that will sense if the horizontal pulse changes in amplitude too much, and will disable the circuits if this happens. Many receivers also have a related type of circuit called an "X-ray protection" circuit that either holds the high voltage down to a safe level or disables it if it should go too high. The reason for this circuit is that if the high voltage is excessive, the electrons may bombard the screen too hard, producing dangerous X-rays.

Focus Voltage

Focus voltage, as mentioned earlier, is derived from the flyback trans-former. Some use a tap, or separate secondary winding in the flyback transformer to provide the appropriate voltage, and a focus rectifier to convert to DC. The voltage can be adjustable using a high voltage pot and high voltage resistors. Note here that ordinary carbon resistors can-not be used for high voltages. The voltage will cause ionization and the resistors could literally explode. Special high voltage resistors are used in these cases. The focus voltage can also be varied using a variable induc-tor. A variable coil resonates with the capacitor at the fundamental fre-quency of 15.75 KHz. Tuning the coil to resonance then decreases the amplitude of the pulse applied to the focus rectifier because the imped-ance of the circuit will be higher. A tuned transformer may also be used to vary the focus voltage. Varying the coupling of the transformer varies the amplitude of the pulse applied to the focus rectifier, thereby adjusting the output voltage. In most modern receivers, the same high voltage that supplies the CRT is applied to a special voltage divider made of high voltage resistors and sealed in high voltage insulation. This device is of-ten referred to as a focus block, as shown in *Figure 9-43*. Often, the focus block also contains additional resistors to divide the voltage to a lower value needed by the CRT screen grids. A second potentiometer is used to adjust the screen voltage.

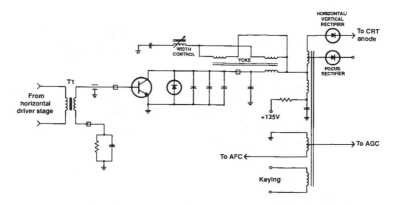

Figure 9-42. *A simplified schematic diagram of an HV/horizontal output section.*

Reducing the Incidence of Failure

To improve the failure rate of the high voltage and focus circuits, many receivers manufactured in the 1970s began to use lower voltage flyback transformers, followed by a solid state tripler. The output of the flyback might be 10KV or so, and the tripler produces around 30KV at its output for the CRT. The tripler usually had a tap within it at a lower voltage to use as the focus voltage. The solid state tripler was found to have a high failure rate. Manufacturing quality control improved dramatically in the 1980s, and now most receivers no longer use a tripler. The flyback transformers produce the full output voltage and have a solid-state high voltage rectifier built in to them. These units are sometimes referred to as integrated flyback transformers, or IFTs. Some even have the voltage divider for the focus voltage and the focus control built into them. Also in the 1970s, some receivers began to use SCRs, GCSs or SCSs in an effort to improve the efficiency of the horizontal output stage and reduce the failure rate. Once again, this has been abandoned and modern receivers are using bipolar transistors as the horizontal output.

It's a Marvel

Modern television sets are the culmination of decades of advances in electronics components and innovations in circuit design. They provide a high-

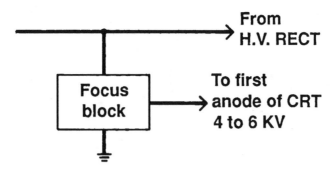

Figure 9-43. In most modern receivers, the same high voltage that supplies the CRT is applied to a special voltage divider made of high voltage resistors and sealed in high-voltage insulation: the focus block.

quality picture and are exceptionally reliable. Nevertheless, even the most modern of TV sets do malfunction, and because of their complexity they frequently present a stimulating challenge to the technicians who are called upon to service them. An understanding of the individual circuits and their interactions in one of today's TV sets, can help make the service process more efficient.

Chapter Ten

Electronic Tuner Theory and Troubleshooting
By Steve Babbert

The advent of the electronic tuner for television sets marked a radical departure from the existing technology. It eliminated the need for most of the moving parts used in the mechanical tuner. Numerous switch contacts, shafts and gears, as well as many of the inductors and capacitors, were replaced with microprocessors, PLLs (phase lock loops), prescalers and varactor diodes.

The mechanical tuners often developed problems due to failure of one or more switch contacts. Sometimes these problems could be solved by cleaning. Other times rebuilding or replacing the tuner was the only solution. In any event, the problems were usually easy to diagnose and understand.

When troubleshooting the electronic tuner, which bears very little resemblance to the mechanical tuner, many technicians find it difficult to pinpoint the source of a problem. This article will attempt to demystify the electronic tuner and outline some basic troubleshooting procedures. But first, let's look at exactly what the tuner does.

The Purpose of the Tuner

The purpose of the tuner is to select a single radio frequency (RF) channel from the many channels that make up the UHF or VHF band, and convert it to an intermediate frequency (IF). While the RF frequency is different for every channel, the IF frequency is always the same. Using this method, all stages of amplification following the tuner won't need to be retuned during channel selection. Conversion of RF to IF is made possible by a process known as heterodyning.

In the heterodyning process, the signal from a local oscillator (LO) in the set is mixed with or "beat" against the incoming RF signal in a stage known as a mixer. (See *Figure 10-1.*) The actual mixing takes place in a nonlinear device. In a linear device such as a resistor, changes in voltage across the device are accompanied by directly proportionate changes in current through the device, so no heterodyning can take place.

Most semiconductors are nonlinear. In modern VHF tuners the mixer is often a MOSFET while UHF tuners generally use a Schottky barrier diode. In the early days of radio this stage was often referred to as the "first detector." This term was supplanted by the commonly-used "mixer" many years ago, reserving the use of the term "detector" for the stage that extracts the so-called intelligence or baseband audio, video or both from the carrier after amplification in the IF stages.

Mixer Output

The output of the mixer contains the sum and the difference of the two input frequencies and in some cases the two input frequencies as well. (This is determined by what type of mixer is used.) Mixing 113 MHz (the channel 4 LO frequency) with 69 MHz (the center of the channel 4 RF frequency) will result in a 44 MHz and a 182 MHz output.

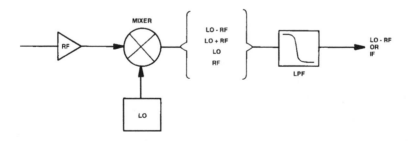

Figure 10-1. *In the heterodyning process, the signal from a local oscillator (LO) in the set is mixed with or "beat" against the incoming RF signal in a stage known as a mixer.*

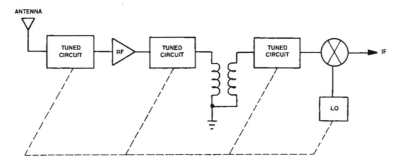

Figure 10-2. *The electronic tuner achieves the same results by varying the DC voltage on a series of reverse-biased varactor diodes that are used as frequency determining components in the above mentioned stages.*

An in-depth discussion of the various types of oscillators and mixers is beyond the scope of this chapter. Suffice it to say that after the filtering of any undesired remnants at the output of the mixer, only the IF frequency will remain.

In the NTSC system, the difference frequency, which is centered at 44 MHz, is used for the IF. The actual frequencies of the sound and video carriers are 41.25 MHz and 45.75 MHz, respectively.

Any channel in the UHF or VHF band can be converted to the IF frequency by selection of the appropriate LO frequency. In the most common mechanical VHF tuners (rotary style), when the channel is changed, a different set of inductors and/or capacitors is switched into the LO circuit. These are frequency determining components and therefore change the LO frequency. The UHF tuner generally uses a set of continuously variable capacitors for tuning.

The first RF amp, which is often referred to as a preselector since it is ahead of the mixer, is also tuned at its input and output by this method to select an individual channel. The input to the mixer also uses this tuning method. The electronic tuner achieves the same results by varying the DC voltage on a series of reverse-biased varactor diodes that are used as frequency determining components in the above mentioned stages. (See *Figure 10-2.*)

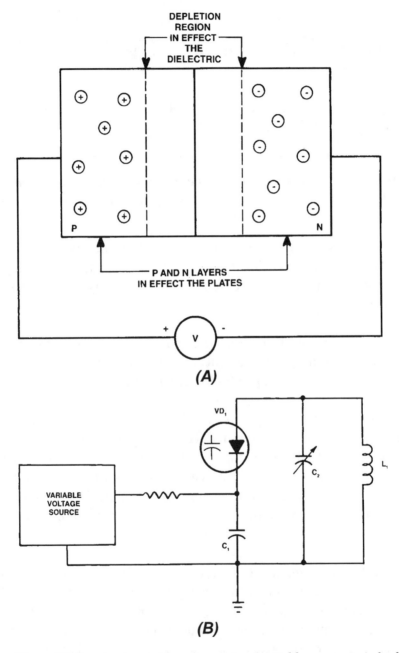

Figure 10-3. A reverse-biased varactor (variable reactance) diode exhibits capacitance which is inversely proportional to the voltage across its P-N junction.

The Varactor Diode

A reverse-biased varactor (variable reactance) diode exhibits capacitance that is inversely proportional to the voltage across its P-N junction. The greater the reverse bias, the lower the capacitance; hence the higher tuned-circuit resonant frequency. (See *Figure 10-3A*). Physically, the change in capacitance is attributable to the change in the width of the depletion region under varying reverse bias conditions.

In semiconductor diodes, a depletion region or space-charge layer exists at the junction of the P- and N-type semiconductor layers. This is a result of mobile charge carriers drifting from the P layer to the N layer and vice-versa. Under no-bias conditions, this diffusion flow reaches equilibrium at some point. Since the depletion region is devoid of charge carriers, it acts as a dielectric. (In this case, the insulating material between the two plates of a capacitor.)

The opposing surfaces of the P and N layers effectively act as the plates of the capacitance. As the reverse bias is increased, the depletion region widens, which reduces the capacitance. (Capacitance is inversely proportional to the distance between the plates.) Though many kinds of diodes exhibit this effect, by controlling the doping profile at the junction during manufacturing, the varactor diode can be tailored to a specific application. Varicap and Selicap are two tradenames for the varactor.

Adjusting the Tuning Voltage

The simplest electronic tuners that are often found in older TVs and VCRs, use a series of potentiometers (presets) to adjust the "tune voltage." A different potentiometer and band selector switch combination is connected into the circuit each time a new channel is selected via a push-button. Some battery operated TVs use a single potentiometer that is adjusted to select a particular channel. This design still uses a band selection switch. Most TVs and VCRs, however, use some form of voltage synthesizer to supply the tune voltage. (See *Figure 10-3B*).

In newer electronic tuners, one or more of the components that comprise the voltage synthesizer may be contained within the tuner housing. (The prescaler is often part of the tuner assembly.) This can place some constraints on what the technician can do.

This isn't to say that it is impossible to troubleshoot inside of the tuner housing. However, if the electronic tuner itself develops a problem it is generally replaced as a unit. The ancillary circuits, however, often can be serviced by the technician.

The Voltage Synthesizer

I recently repaired the synthesizer board and varactor tuner in a Sanyo Model 91C94N. In this chassis, all components related to voltage synthesis are located outside of the tuner. The main blocks that will be covered are the prescaler, the PLL, the microprocessor, the tuner band selector switch and the low-pass filter. (*Figure 10-4.*)

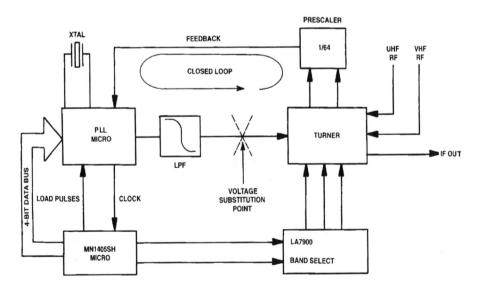

Figure 10-4. The main blocks of the voltage synthesizer include the prescaler, the PLL, the microprocessor, the tuner band selector switch, and the low-pass filter.

The Band Selector Switch

In the electronic tuner, some sections are common to VL (low band VHF), VH (high band VHF) and UHF while other sections are used exclusively for one of these bands. Tuners incorporating a superband section follow this rule as well. Tuning of these bands is done separately because the varactor diode has a limited capacitance range and therefore cannot cover the entire tuning range of the VHF band (54 to 216 MHz which is a 4:1 tuning ratio) much less the combined UHF and VHF band.

The purpose of the band selector switch is to enable the appropriate sections of the tuner as needed when a channel is selected. In this chassis, the band selector switch IC, an LA7900, is controlled by the microprocessor, an MN1405SH, via two lines. The tuner is controlled by the band selector switch via three lines.

When a channel is selected, the microprocessor, in response to an internal program, instructs the band selector switch to open or close one or more of its three internal switches. Switching diodes within the tuner either pass or block signals depending on whether they are forward or reverse biased by voltages from the band selector switch. The result of this switching scheme is that only the sections of the tuner that were designed to accommodate a given band will be activated. In this band selector switch, one switch places the tuner into the UHF mode, one places it into the VHF mode and the remaining switch selects between VL and VH when the tuner is in the VHF mode. (See *Figure 10-5*.)

The tuning circuits must make a relatively far jump between channel 6 (VL's highest) and channel 7 (VH's lowest). The center frequency of channel 6 is 85 MHz and that of channel 7 is 177 MHz. The difference is 92 MHz. Most channels differ by 6 MHz. The reason for this gap is that those frequencies were allotted for other services. For example the FM broadcast band uses 88 to 108 MHz.

The tuned circuits are shifted to a higher frequency by forward biasing a switching diode which shunts a portion of an inductor to ground. (See

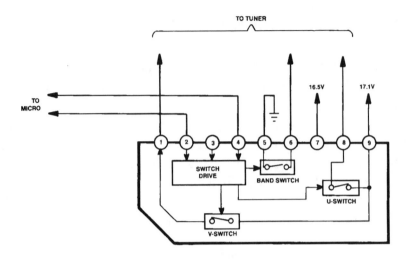

Figure 10-5. *In this band selector switch, one switch places the tuner into the UHF mode, one places it into the VHF mode and the remaining switch selects between VL and VH when the tuner is in the VHF mode.*

Figure 10-6.) This reduces the inductance which raises the resonant frequency. The reactance of C1 is negligible at the frequency of interest so it acts as a short. Its purpose is to isolate the switching voltage from ground.

Since the UHF band is so far removed from the VHF band, it isn't practical to try to tune it on a tuner designed for VHF. As frequencies increase, the problems of RF design become greater. For this reason, most of the UHF stages in the tuner are completely different and separated from the VHF stages. Switching diodes are used to activate or deactivate stages as needed. The main stage, which is shared by the UHF and VHF sections, is the VHF mixer. Once the UHF frequency is converted to the IF it is passed through the VHF mixer which in this case acts as an IF amp.

The Tune Voltage

Now that band selection has been covered, let's look at individual channel tuning. The tune voltage measured at the output of the low-pass filter (LPF) ranges from a little over 1V to about 24V. There is some overlapping of voltages between bands. The tune voltage when in the VL band

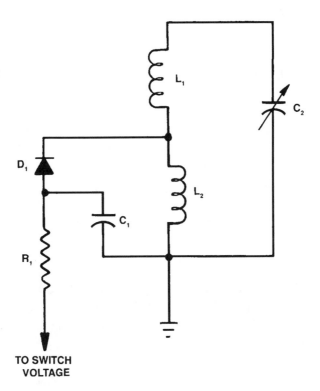

Figure 10-6. *The tuned circuits are shifted to a higher frequency by forward biasing a switching diode which shunts a portion of an inductor to ground.*

will range from 2V to about 20V. In the VH band the range will be from about 7V to 20V. In the UHF band it will range from about 2V to 24V. These ranges may be different for other tuners. Generally the tune voltage will not exceed 28V.

If the tune voltage line (which will be referred to as VT in this chapter) is monitored, it can be seen that the voltage will increase each time a higher channel is selected. Once a higher band is passed into, VT will drop to the floor of that band. Passing into a different band causes different tuner sections to be activated so a specific value of VT won't tune the same frequency in different bands. If VT is monitored ahead of the LPF (which in this case provides inversion and amplification) the voltage increments will be smaller while stepping through channels.

The PLL

One of the most important and complex blocks in the voltage synthesis circuit is the phase lock loop (PLL). Basic PLL theory won't be covered here. The PLL receives instructions from the microprocessor and feedback from the tuner itself and outputs an error voltage which becomes the tune voltage. The PLL IC in this chassis is an MN6049. One of the blocks in this IC is a 13-bit programmable divider. This divider receives an input signal from the UHF or VHF LO in the tuner after having been passed through a prescaler and an on-board divide-by-2 counter.

The Prescaler

The prescaler in this chassis is a separate shielded module mounted beside the tuner. It consists basically of a 4020 CMOS 12-stage binary counter IC configured to divide by 64. (See *Figure 10-7*). Its input comes from the VHF LO or the UHF LO (depending on which is active).

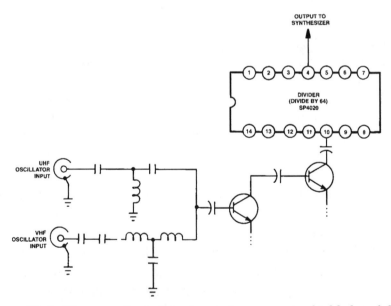

Figure 10-7. The prescaler in this chassis is a separate shielded module that consists basically of a 4020 CMOS 12-stage binary counter IC configured to divide by 64.

The LO signal is buffered and amplified by a two-stage amp once inside the prescaler module before application to the divider. Since the UHF/VHF LO frequencies are related to the tuned frequency, they can be used as feedback by the PLL to determine what frequency the tuner is tuned to at a given time. This feedback signal is essential for proper operation.

Once the feedback signal is divided further by the divide-by-2 block and the programmable divider, it is compared to a stable reference at the phase detector. If the two frequencies don't match, the output of the phase detector, which is basically a voltage comparator, will toggle high or low as needed in an attempt to bring the tuner frequency in line.

Actual locking between the two frequencies never takes place. Instead, the output of the phase detector responds quickly each time the tuned frequency overshoots or undershoots the reference. This holds the tuner frequency within an acceptable tolerance.

The Programmable Divider

The programmable divider receives a specific set of data from the microprocessor for each channel. By selecting the correct divide modulus or division factor, the divider can convert any incoming divided down LO frequency to the reference. This way, no matter what frequency is tuned by the tuner, once the LO frequency for that channel is divided it will be suitable for comparison. Since this system uses feedback from the tuner to create the tune voltage, it must be viewed as a closed loop. As in any loop system, a problem in one section will upset measurements in all sections.

This PLL has an on-board oscillator that uses an external 3.581055 MHz crystal. This signal is divided by 8 before being passed to the load pulse generator. It is also applied to the microprocessor where it is used as a clock signal. This way the microprocessor will be locked to the PLL. Another divider in the PLL divides the signal by 3667, giving 976.5625 Hz before it is applied to the phase detector where it is used as the stable reference.

The Load Pulse Generator

The input to the load pulse generator, at pin 5 of the PLL, which is derived from the microprocessor, provides timing information for the loading routine. At first glance with an oscilloscope, the signal looks like a narrow positive-going pulse with a frequency of about 57 Hz. If the timebase of the scope is adjusted to expand this pulse on the display, it can be seen that the pulse is actually a cluster of four pulses.

Though the PLL uses a 13-bit divider, there are only four data lines coming from the microprocessor to the PLL. The load pulse generator, which receives a signal from the microprocessor as well as a signal from the divide-by-8 divider, working in conjunction with data latch A and four AND gates, provides a means for sequentially loading latches B through E. If any one of the four data lines is scoped at the same time as the load pulse, it can be seen that a 4-bit data burst is present for the duration of the load pulse cluster.

The data burst is different on each of the four data lines. The data will change each time a new channel is selected. Each latch will maintain its status after it is loaded thus acting as a data retainer (or memory). This way the 13-bit word will be available at the data input of the divider during the interval between the load pulses.

The Division Factor

The following example will show how the division factor of the programmable divider is established. The LO frequency for channel 4 is 113 MHz. Once this is divided by 64 in the prescaler and 2 in PLL's on-board counter, the result will be 882,812.5 Hz. We know that after division in the programmable divider it must equal 976.5625 Hz to be equal to the reference. We can find the division factor by dividing the last number into the first which gives 904.

Therefore, in order to tune channel 4, the programmable divider must divide by 904. The binary equivalent of this number is 1110001000. This number is stored in the microprocessor's built-in memory at the "channel 4" address. This is only a 10-bit number and could be handled by a 10-bit divider. However, channel 83, which uses an LO of 971 MHz, requires a divide modulus of 7768, which translates to 1111001011000 (13 bits).

The Phase Detector

The output of the programmable divider, which provides one of the inputs to the phase detector, is available at pin 10. The divided-down reference, which is the other input to the phase detector, is available at pin 9. By scoping these two signals simultaneously, it can be seen if they share the same frequency and phase (which incidentally is the job of the phase detector). In a properly working tuner, this will be the case when any channel is selected. It makes no difference whether the channel is active or inactive.

The most important parameter of the phase detector output at pin 8 is the average DC value. This value is dependent on which channel has been selected. This voltage is applied to the input of a low-pass filter. The LPF removes any AC components that were generated by the phase detector.

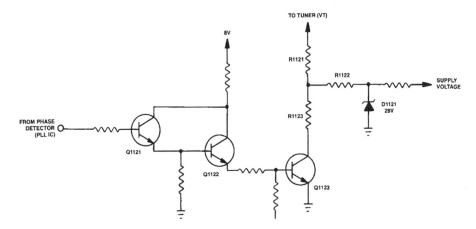

Figure 10-8. The raw tune voltage is taken from a 28V Zener diode D1121. This voltage is applied to a voltage divider consisting of R1122, R1123, and Q1123.

The raw tune voltage is taken from a 28V Zener diode D1121. This voltage is applied to a voltage divider consisting of R1122, R1123, and Q1123. (See *Figure 10-8*). The voltage that is routed to the tuner control input is tapped from this divider by R1121. Transistor Q1123 serves as a variable resistance. It is also an inverter.

When it is turned fully on, VT will be at its lowest point. When it is fully off, VT will be at its maximum. Normally, it will be partially on, holding VT somewhere in between. Notice the LPF is active and uses a Darlington configuration, hence it has a high gain. A change of only a fraction of a volt at the input results in a change of several volts at its output.

Chapter Eleven

Servicing Vertical Foldover Problems
By Homer L. Davidson

Foldover is a form of distortion that can be caused by malfunctions in either vertical or horizontal circuits. Overlapping of the picture at the top or bottom of the TV screen is called vertical foldover. In foldover, one portion of the picture overlaps another portion of the picture, accompanied by distortion. Vertical foldover can occur at the top, middle or bottom of the raster.

The most common form of foldover is insufficient picture height with a half inch or so of foldover in the middle of the screen. When foldover occurs at the top of the picture, it may be accompanied by black and white lines. Excessive height at the bottom of the picture may result in overlapping of the picture or vertical foldover. In some cases, the bottom area of the picture may be raised up, and you will notice a black area at the bottom.

The Cause of Vertical Foldover

Vertical foldover is caused by defective components in the vertical output, bias or feedback circuits. Often, a leaky or open electrolytic capacitor, a change in the resistance of a resistor in the feedback circuits, or a defective transistor or IC component is the cause of foldover. In older TVs a transistor or leaky coupling capacitor to the vertical yoke winding may be the cause of vertical foldover. Feedback resistors or electrolytic capacitors can also cause vertical foldover.

Accurate voltage measurements and oscilloscope tests can help locate the defective component. Excessive vertical foldover cannot be corrected by adjusting the vertical height and linearity controls.

If you encounter a TV set that has a vertical foldover problem, check the vertical output transistors to see if they are leaky, and check the bias resistors to see whether their value is significantly different than the specified values. Check to see if the voltages on the bottom vertical output transistor are according to the specifications.

It may be necessary to replace both vertical output transistors in order to solve a foldover problem. Although the oscilloscope cannot pinpoint the defective component, observation of the input and output waveform can help the technician determine if normal sweep is present. (*Figure 11-1.*)

Injection of a normal waveform into the input terminal of the vertical output IC can help determine if the foldover problem is caused by a defective IC or problems in the output feedback circuits. Vertical linearity or foldover problems can be caused by vertical feedback and bias circuits.

Figure 11-1. If the problem is vertical foldover, check the waveforms at the input of the vertical deflection IC, and the output of the deflection IC to the vertical yoke winding.

A suitable diagnostic procedure is to shunt all electrolytic capacitors in the output and feedback circuits one at a time. If shunting these electrolytic capacitors does not produce any improvement in the problem, you can be almost certain that it will be necessary to replace the vertical output IC to solve the foldover problem.

Careful resistance measurements in the bias and feedback circuits can turn up a leaky capacitor or resistor that may be causing foldover. Remove one end of the resistor to isolate it from the remainder of the circuitry for accurate tests. In many cases you will find that the defective resistor has increased in resistance.

Transistor Output Foldover Problems

In early TV chassis with transistors in the vertical output circuits, vertical foldover may be caused by a defective transistor, an electrolytic coupling capacitor, bias diodes, or a change in resistance of the feedback resistor. In many cases measurements will show that the voltages at the terminals of the vertical output transistors are incorrect when the TV set exhibits foldover.

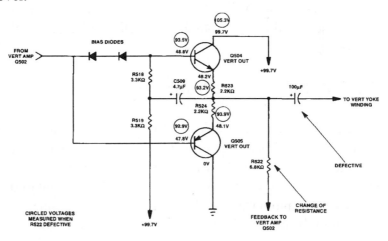

Figure 11-3. When vertical foldover was the problem in a Sharp C-1551 set, the replacement of the 100mF yoke coupling capacitor and feedback resistor R522 (6.8K) restored normal operation.

In a Sharp C-1551 portable with a foldover line at the top of the raster, full vertical sweep could not be obtained. (*Figure 11-2.*) In this case, the voltages at the terminals of the top output transistor (Q504) had increased by 5.6V. Voltages at the bases of both output transistors were almost twice their specified values. At first, I suspected that Q504 was leaky, since voltages on all three terminals of that transistor were close to the same value. When tested in-circuit, Q504 appeared normal. I decided to remove the transistor for out-of-circuit tests. The transistor appeared normal out of circuit. These tests are not conclusive, because a transistor may be intermittent under load, or its characteristics can change when heat is applied to remove it from the circuit.

Because the symptom pointed to Q504, even though it seemed to test good, I replaced this 2SC1448A transistor with an SK3054 universal transistor. The symptom remained. I checked all bias resistors, and they were normal. A measurement of the resistance of feedback resistor R522 (6.8KΩ) in the circuit revealed that this value was higher than normal. I disconnected one end of R522 in order to measure it free from the influence of other resistances that might be paralleling it. The value of this resistor had increased to 15.58KΩ, so I replaced it. I replaced the vertical coupling capacitor as well, since in some sets has caused foldover problems. This cured the vertical foldover problem at the top of the picture. (*Figure 11-3.*)

One-Half-Inch Foldover Problem

The picture on the screen of a General Electric 19PCF set had shrunk to a very small height. When the vertical hold control was adjusted, the picture expanded to two inches with a half-inch foldover in the center. Readjustment of the height control did not increase the height of the raster. Because the symptom was insufficient height, both output transistors were checked in the circuit (*Figure 11-4*), but both tested good.

A measurement on Q601 indicated high collector voltage (85V). Usually this voltage measures around 65V, measured at the 116V source through R647 (1.5KΩ). The voltage between the base and emitter on both tran-

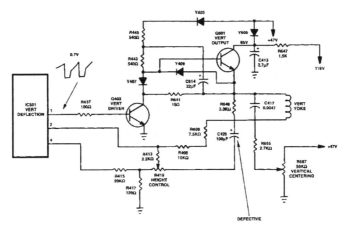

Figure 11-3. *When a General Electric 19PCF chassis exhibited a vertical problem, only one-half inches foldover in the middle of the screen could be varied with the vertical hold control.*

sistors measured 0.6V. Since these transistors are NPN types, these forward-bias measurements indicated that both transistors were normal. A quick look at the IC waveform at pin 1 of IC501 with the oscilloscope showed that this was normal at 0.7V, indicating that the foldover problem must be caused by a malfunction in the vertical output circuits.

Figure 11-4. *To locate the vertical deflection circuits in a Sharp 19J63 or 19J65 chassis, look for IC501 on its metal heat sink.*

Since a defective electrolytic capacitor or a change in feedback resistors can also cause foldover, I shunted each of the capacitors, one at a time, by turning off the set, clipping a new capacitor across the suspected one in the set, then turning on the set again to observe the results. When I clipped a new 22μF capacitor across capacitor C614, there was no change in the picture. When C425 was shunted with a 100μF electrolytic capacitor, however, the vertical sweep returned to full deflection, and the center foldover problems had disappeared.

Sharp 19J63 and 19J65 Foldover Problems

When you encounter a TV set with a vertical sweep problem, always check the vertical output waveform from the vertical deflection IC, to determine if the cause of the problem is in the output circuits. In many cases improper vertical sweep is accompanied by foldover symptoms.

Locate the vertical deflection IC (IC601) and output IC (IC501). (*Figure 11-5.*) If the input waveform appears to be near normal, go directly to the vertical output. If the vertical output waveform at pin 2 of IC501 is too low, it may indicate insufficient vertical sweep.

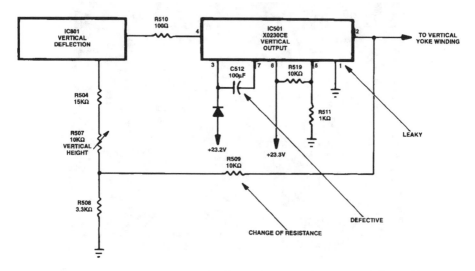

Figure 11-5. Components C512 (100μF), R509, and IC501 can cause foldover in the Sharp 19J63 and 65 chassis.

Carefully measure voltages and resistances at the terminals of the suspected IC before attempting to remove and replace it. Test all components tied to each IC terminal. Low supply voltage at pin 6 can indicate a leaky IC501 or improper low voltage source. Since the IC is available only through the manufacturer's service depot or factory channels, take extra pains to be sure that the IC is defective before you start to desolder it. IC501 has been found to be leaky in both the 63 and 65 chassis.

Before removing the suspected IC501, shunt capacitor C812 (100µF) with a known good electrolytic capacitor. (*Figure 11-6.*) To perform this operation, turn the set off and clip a new capacitor across the old one, then turn the set back on. If C812 is open, the raster will return to normal. For further confirmation, measure the resistance between pins 3 and 7 to determine if C812 is leaky. Do not overlook the possibility that the problem may be caused by feedback resistor, R509. This resistor can change resistance and cause vertical foldover.

If careful observation of waveforms and voltage and resistance tests at the outputs of IC501 don't demonstrate conclusively that that particular

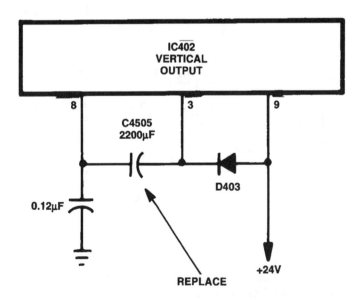

Figure 11-6. A 220µF electrolytic capacitor located near the vertical output IC, IC402, in an RCA CTC177 caused vertical foldover problems.

IC is the cause of the problem, shunt all electrolytic capacitors in the vertical output circuits one at a time. Also, remove one end of R509 (10K) for correct resistance measurement. If this resistance reading is within specification, and if the insufficient sweep and foldover problems still exist after shunting each capacitor, remove and replace IC501.

RCA CTC177 Foldover Problems

In the RCA CTC177 chassis, a faulty C4504 may be the cause of insufficient vertical sweep and foldover. Another symptom that may occur when this capacitor becomes faulty is an intermittent white line accompanied by intermittent vertical deflection.

Vertical output IC402 runs quite warm in several of the latest RCA chassis. Heat from this IC may cause C4505 to swell up, bulge, or have a blown top, since it is mounted in close proximity. (*Figure 11-7.*) Gas inside the capacitor is usually the cause of such physical problems, and heat may accelerate the process.

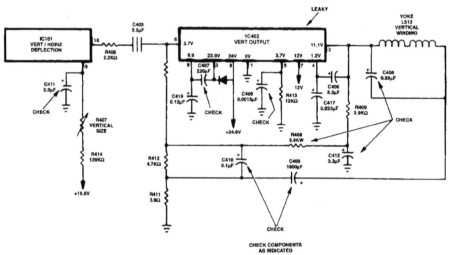

Figure 11-7. If you encounter a vertical foldover problem in a Panasonic PC-11T31R chassis, check the components in the vertical circuits pointed out here for poor linearity and foldover problems.

Although the heat from the output IC may not damage C4505, check this capacitor any time you encounter the different vertical symptoms. If this 220μF capacitor appears to be deteriorating physically, remove it from the PC board and replace it with one that has longer leads. Fold the electrolytic capacitor away from the hot output IC. Check this capacitor in several of the latest RCA chassis for vertical problems.

Panasonic PC-11T31R Chassis Foldover

In the Panasonic PC-11T31R TV set, IC101 contains the vertical and horizontal deflection circuits, with a vertical preamp output at pin 10. The vertical size control (R407) varies the voltage on pin 9 of IC101. Check the sawtooth waveform on pin 10 to determine if the vertical oscillator and preamp circuits are functioning. A 1.5V sawtooth pulse is sufficient to drive the vertical output IC, IC402. (*Figure 11-8.*)

Figure 11-8. When you don't have a schematic diagram available and you want to locate the vertical output IC, look for an IC mounted on a large heat sink.

The sawtooth waveform from pin 10 is passed through R406 and C403 to the input terminal of IC402, pin 6. IC402 amplifies the vertical signal and provides a pulse to sweep the vertical winding. Notice that the vertical yoke winding connects directly to the IC output, pin 2, and the return winding goes through C408 (1000mF) to R411 (3.9+), to common ground.

Two separate voltage sources are fed to this IC, one to pin 9 (24V) and one to pin 7 (12V) to power IC402. If the voltage of either (or both) of these sources is not correct, the symptom may be insufficient vertical sweep and foldover. A faulty IC402 may also cause this problem. If you encounter insufficient sweep and foldover on one of these sets, check the voltages on pins 3 and 8 for a possible leaky capacitor C407 (220mF).

Possible causes of improper vertical linearity and foldover in these sets are: IC402, R408, C409, C408, C410, C412, C407, C406, and C411. Shunt each capacitor in turn with a known-good capacitor to determine if it is open. If this process causes you to suspect one of the capacitors, carefully measure the resistance across its terminals to confirm whether it is leaky.

Locating Defective Vertical Components Without a Schematic

When you're faced with a set in which the symptom points to a problem in the vertical circuits, and you don't have a schematic readily available to help you find the vertical section so you can test the components, a careful examination of the circuitry in the set may help you locate the vertical section. Look for a vertical output IC on a large heat sink.

Once you have located the vertical circuits, carefully measure voltages at the terminals of the vertical output IC, and examine waveforms at the IC terminals with the oscilloscope. Check the waveform at the IC terminal that connects to the yoke winding. Start with the yoke and trace the PC wiring back to the output pin. Improper waveform at this output pin may indicate a defective IC, improper voltage sources, or a defective component tied to each pin terminal.

Next check for one or two voltage sources feeding the vertical IC from the low-voltage power supply. The highest voltage is the voltage supply pin. A 24V to 25V source powers the vertical output IC component. Scope the input terminal which is usually a sawtooth or negative pulse (0.9V to 1.5V).

You can signal trace the vertical circuits with one of the latest TV schematics, which includes a vertical output IC. Most of the latest vertical IC output circuits are just about the same and can be used to signal trace, observe waveforms and measure voltages.

Notice that the vertical yoke winding ties directly to pin 4, with the return winding through a 680μF electrolytic capacitor and a 3Ω resistor to common ground. Besides the 680μF capacitor, check 2.3μF, 1μF, 0.033μF, 100μF, and 470μF capacitors any time you encounter poor vertical linearity and foldover problems.

Conclusion

Whenever the problem is vertical foldover, carefully observe waveforms into and out of the vertical output IC to determine if the IC is defective, or if the problem is caused by other components that are defective. Remember vertical foldover is caused by problems that occur in the vertical output, feedback and bias circuits.

Shunt electrolytic capacitors in the vertical output and feedback circuits to check for open conditions. Carefully measure resistances across electrolytic capacitors to determine if capacitors are leaky. Measure the voltage source to supply pin on output IC to determine if the voltage is proper.

Chapter Twelve

Constructing a Tuner Subber

By Dale C. Shackelford

In addition to replacing defective components, cleaning dirty potentiometers and aligning coils, many consumer electronic servicing technicians use their skills and imagination to design and construct test equipment from spare parts which are often destined for the discard pile. One of the easiest and least expensive pieces of test equipment to construct is the tuner subber.

While a tuner subber is not in the same league as a power supply project, insofar as necessity is concerned, tuner subbers can save several man-hours in the diagnosis of a television receiver with signal processing problems, by verifying or absolving the tuner as the basis of a problem. With the investment of an hour of labor and less than $10.00, any competent technician can construct a tuner subber that will provide years of service.

Obtaining the Tuner

The main component of any tuner subber is, of course, the tuner. While almost any television tuner may be used as the cornerstone of the project, it is best to use a unit salvaged from a discarded color receiver rather than one from a monochrome set, because some monochrome tuners may employ components or filtering circuits to suppress certain signals used in color sets in an effort to reduce interference. Additionally, some monochrome tuners lack the ability to properly tune or process certain frequencies needed in the color receiver.

You also have a choice between analog or digital tuners for this project. While either type of tuner will provide a signal sufficient to check the receiver, the use of an analog tuner on a digital receiver will sometimes result in a garbled or distorted output. Analog tuners are much easier to convert to subber use than their digital descendants, as small variations in source voltages do not affect the overall signal processing capabilities of the analog unit to the extent that they affect the digital package. Also, electronic tuners usually require several conditioned sources to operate, while the analog version requires only two: B+ and gain.

Before deciding upon a particular tuner, it is wise to determine whether or not a schematic diagram or other documentation is available for the particular model to give you some idea as to the voltage requirements. An alternative is to find a working receiver with a similar tuner, and to measure the working voltages at the B+ and AGC connections.

While the B+ voltage should remain constant, the AGC voltage may vary somewhat with the relative signal strength. Checking the specifications on the schematic, or adjusting the AGC control on the working set should give you some indication of the AGC parameters. As a general rule, analog tuners require about 10V DC at the B+ connection, while AGC voltages will vary between 1V DC and 3V DC.

Retrieving the Tuner

Many television sets have their tuners (both VHF and UHF) mounted on a single plastic or metal frame (*Figure 12-1*), which is in turn mounted inside the cabinet of the receiver. This configuration allows both tuners to be removed from the set as a single unit after all connections (IF cables, power supplies, etc.) to the main circuit board have been disconnected. Leaving enough wire on the tuner(s) means that new wires will not have to be soldered to the tuner, where excessive heat or accidentally dropped solder could damage sensitive coils or other components.

In some instances, the frames to which the tuners are mounted are also used by the manufacturer to mount potentiometers for color, brightness, volume and other controls. In the event the unit you have decided to use does not have a pot mounted in it, simply drill a hole through the bracket in a space big enough to accommodate a pot, making sure that the positioning of the pot does not interfere with the operation of the tuner. This

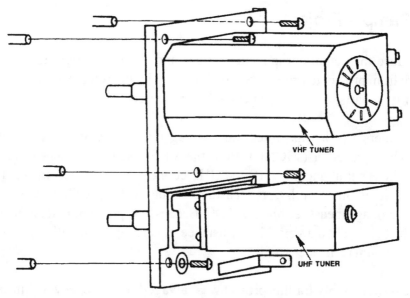

Figure 12-1. Many television sets have their tuners (both VHF and UHF) mounted on a single plastic or metal frame, which is in turn mounted inside the cabinet of the receiver.

component will be used to manually adjust the gain to the tuner subber in lieu of the automatic gain control circuit. In this application, we will be using a 5KΩ potentiometer, but the physical size is not important.

The Power Supply

The next consideration is the power supply. While almost any power supply with a DC output of at least 15V will suffice, the key to the project is simplicity and low cost. The easiest power supply to obtain, and the least expensive, is one salvaged from a portable stereo receiver. Often, these devices use fairly efficient power supplies which are often self-contained, meaning that the transformer, rectifier circuit and AC input connections are all mounted on a single circuit board. This configuration allows the power supply to be removed as a unit, while making it very simple to use as a power source for the tuner subber. A few of these self-contained power supplies also feature variable resistors, which allows the DC output to be adjusted.

A Sample Project

Because the actual components, values and wiring configurations will vary from project to project, based upon technician preferences and availability of parts, a sample project will be discussed.

The tuners used in this project were salvaged from a scrapped KTV color television receiver (13CNR). Both the VHF and UHF tuners were removed as a unit and all VHF and UHF connections were left intact. The B+ voltage on a similar set measured about 9V DC, while the average AGC voltage measured about 2V DC. Because the potentiometer mounted in the bracket had a very high resistance, it was replaced with a pot with a maximum resistance of 5KΩ.

The power supply for this project was salvaged from the recycle bin, and looked to be from a Magnavox portable stereo receiver, the case of which had long since been discarded. The case for the tuner subber was constructed from pieces of 1/8th-inch plywood that had been laying around the shop for years, knowing that someday, someone would put it to good use. Had I not found the plywood, I had planned to use an old plastic television cabinet to construct the box. Since the plywood was already cut to a fairly uniform size, a minimum amount of trimming with a razor knife resulted in an 8-inch by 8-inch by 8-inch plywood box, more than enough to contain the tuners, transformers, rectifier board and related hardware.

Figure 12-2 depicts the schematic of the sample project tuner subber which was built entirely from discarded parts found around the service center, and took less than an hour after the design was finalized. Because the power supply is more or less typical, it is shown in block form, though the voltages pertinent to the project are indicated. Again, technician preferences and available components may change the overall design of the project. Some components may be added and others changed, while others might be deleted altogether.

Figure 12-3 shows the various power and signal inputs for the project.

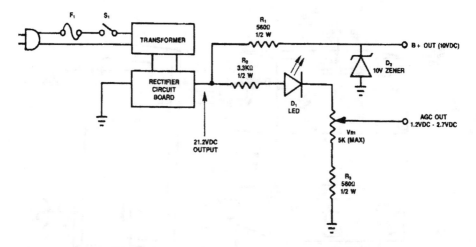

Figure 12-2. The schematic of this sample project tuner subber was built entirely from discarded parts found around the service center. Because the power supply is more or less typical, it is shown in block form, though the voltages pertinent to the project are indicated.

Overcoming a Problem

The only head-scratcher encountered in this project was that the B+ would vary about 2V when I adjusted the gain control VR1. While the variation had no visible effect upon the operation of the tuner subber, it was obvious that there was a defect in the design of the circuit. After a little thought, I placed Z1 in the circuit, which held the B+ voltage at a steady 10V DC while the gain control could be varied from 1.21V DC to 2.71V DC. Had the tuner required a B+ source voltage of a different value, the value of the zener diode would simply have been changed to accommodate the requirement.

Obviously, D1 could have been replaced with a general-purpose rectifier diode, but the LED used in the circuit is not only functional, but is also used as a power-on indicator when mounted through the front of the cabinet. This LED also allows the capacitors in the rectifier circuit to completely discharge once the unit has been turned off. In this project, SW1 is ganged with VR1.

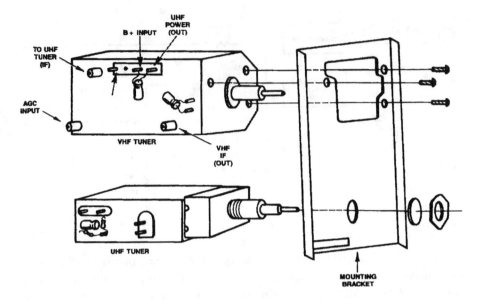

Figure 12-3. The various power and signal inputs for the project.

Antenna/cable inputs, which can also be salvaged from a scrapped television along with associated isolation circuitry, can be mounted in the side
or back of the project box. Make sure to use shielded coaxial cable or
antenna ribbon wire to connect the antenna/cable inputs to the tuner.
Additionally, the IF output cable can consist of a length or RG59 coaxial
cable soldered on one end of the IF output terminals of the tuner, while
small alligator type clips (or whatever connectors are desired) are soldered to the other.

Using the Subber

Once finished, you will have constructed a piece of test equipment which,
on average, would have cost a couple of hundred dollars new. While your
project subber may not take the place of a currently used off-the-shelf
unit that you may now be using, it's good to know that you have a backup
if the need ever arises. Hobbyists who do not own a tuner subber will find
that they can receive many hours of service from this inexpensive project
unit, while having the satisfaction of building the unit themselves.

Chapter Thirteen

Build This Tester for Infrared Remote Controls

By Ricky Hall

Most technicians use a credit card-size tester to test infrared remote control transmitters. This is a small white card that gives off a dim reddish light when it is illuminated with infrared light. In order to test a remote control unit, the user aims the remote transmitter at the card and presses a button. The user must be in subdued light, and must hold the remote close to the card in order to see the light it emits.

The light from these cards is difficult to see. This, coupled with the fact that these cards are small and easy to misplace, prompted me to find another method of testing remote transmitters.

Using a Remote VCR Controller as a Tester

Some time ago, I hit upon the idea of using a store-bought device to do this job. This unit, called XTRA Link, allows the user to control a VCR from another room. The box at which the user aims the remote transmitter has an LED that flashes when the IR signal is being received from the remote unit.

With this tester, I can get an indication of IR transmitter function from across the room, under normal light. This unit cost about $70. I used it for years and wondered how other TV technicians could do without one.

Building a Remote Control Transmitter Tester from Reclaimed Parts

Several times while I was troubleshooting the infrared remote control transmitter of a TV or VCR, I wondered if it might be possible to use the

IR receiver from a remote-controlled product to make a tester. Every consumer electronics service center, including mine, has discarded TVs and VCRs in storage. Many of these units are remote control. Most of them have a small IR receiver with connections as shown in *Figure 13-1*: +DC V, output and ground. A simple circuit, as shown in *Figure 13-2*, can put an IR receiver to work as an IR remote transmitter tester.

One day while doing some experimenting with an IR receiver from a TV, I hooked the cathode of an LED to the output and the anode to the +9V DC, and it worked. The LED was bright and it lit up or flashed depending on the output from the transmitter under test.

This tester works even better than the XTRA Link, which would occasionally give false indications under certain light conditions.

Technicians can find all the parts to build a tester (except a box to put it in and the 9V supply) inside old TVs and VCRs. I used an old TV antenna power supply box to put it in, and I built two units for my service center.

One of these units is powered with a 9V battery so I can take it on calls. The other unit is supplied by a 9V DC power adapter from an old Atari video game. I put the unit that's powered by the adapter on a shelf, and it stays on all of the time. Whenever I want to check and see if an IR remote tester has an output, I just "shoot" it at the IR tester on the shelf.

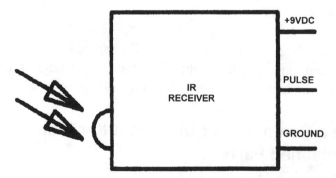

Figure 13-1. If you have a discarded TV or VCR waiting to be scrapped, you may be able to reclaim the IR receiver. Many of them have connections as shown here. IR receivers are also available inexpensively at stores that sell electronics components.

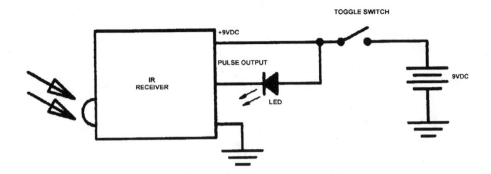

Figure 13-2. A simple circuit, as shown here, can put an IR receiver to work as an IR remote transmitter tester.

Building a Tester from Store-Bought Parts

You could also do what I did and build yourself a nice unit with parts from an electronics distributor. I gave the one that I built to my brother, who is also a technician, for Christmas. I used one of those super bright red LEDs, and it looked store-bought. *Table 13-1* is the list of parts from Radio Shack.

Evaluating the Testers

I find that the IR receivers from TVs work best and are much brighter than the ones used from VCRs. The IR receiver that I built from the Radio Shack parts works super.

PART	NUMBER	APPROX. PRICE
IR Receiver	276-137	$3.49
High Brightness LED	276-066	$1.19
Case (Box)	270-293	$3.99
9V Battery Clip	270-325	$0.26
SPST Toggle Switch	275-624	$2.29

Table 13-1. You can build the IR tester circuit in Figure 13-1, using the bill of materials shown here. The case has a compartment with a door that you remove to insert a 9V battery; a nice feature.

Chapter Fourteen
Homemade Isolation Transformer to Cure H-K Shorts
By R.D. Redden

A bright red, green or blue screen with retrace lines often means a heater-to-cathode (H-K) short in the CRT. When TVs used 60 Hz to power the CRT filament (heater), technicians bypassed this problem by installing a large, heavy isolation transformer for the filament.

Now that filaments are powered by the high-voltage transformer, which operates at a much higher frequency, isolation transformers are small and light. And it's easy to wind your own. The materials for the ones I make cost me under two dollars, and I can wind one in about five minutes.

Actually, the first isolation transformers I used were free. Some TV sets used toroid transformers of about 1-1/2 to 2 inches in diameter near the AC input as a line choke. I snipped these out of junked sets and they worked well as isolation transformers. But as happens with most good freebies, they seem to have become scarce.

Not all toroid cores are the same. The core material affects the permeability and thus the inductance for the same number of turns on similar sized cores.

I wanted a small, light core that would require few turns of wire. Ocean State Electronics (800-866-6626) has a wide selection of cores. Their stock number FT82-75 has a permeability of 5,000, allowing 14 turns of No. 22 wire to produce about 500H to 550H of inductance.

I use GC hookup wire, which has thick insulation. Such thick insulation may not be necessary, but I wanted to be sure a CRT arc would not cause a short in my isolation transformer.

Wiring the transformer is easy. Just cut two 18-inch lengths of the No. 22 hookup wire. You can use the same color and mark one wire for the primary, but a different color for each wire is recommended. Stretch the two wires side by side and slide a core to about the center of the two wires. Make fairly tight turns around the core, alternately doing two or three turns on each side of the starting point. (All doubled turns are the same direction, but starting in the middle of the wire means less wire to thread through the core.)

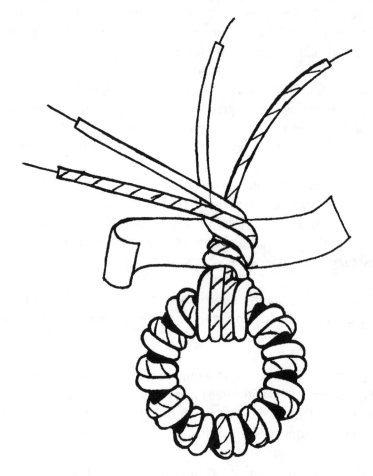

Figure 14-1. *If you encounter a TV set with a heater-to-cathode (H-K) short, an isolation transformer such as this one, which you can make yourself, will allow you to restore the set to almost perfect operation.*

You should be able to just fit 12 turns on the core in a single layer. Then make one more turn with each set of wires in a second layer for a total of 14 turns. Give the ends a twist next to the core and wrap a little tape around the twist to hold it together. You should have about two inches of wire left over for external hookup. (See *Figure 14-1.*)

I needed to answer three questions about the transformer before using it. Did it cause an excessive load on the HVT? Did it supply enough current for the filaments? Would it cause distortion of the TV's picture?

Since the voltage going to the primary of the isolation transformer is shaped like the retrace pulse that is its source, calculating the current drawn by the isolation transformer with its secondary open (no load current) would be difficult.

But by leaving the secondary open and hooking the primary to the filament supply from the HVT of a used 19-inch set, I could monitor the B+ current to the horizontal output transistor and check for the additional load current caused by the isolation transformer. I found the current increased from 0.315A to 0.316A, or 0.317A (the last meter digit fluctuated). This increase of only 1mA to 2mA did not seem excessive.

Still monitoring the horizontal output current, but with a test pattern on the TV and the contrast and brightness turned to maximum, I got a reading of 0.625A, whether the filament was powered directly or through the isolation transformer.

I was not able to see any difference in the brightness or contrast of the picture whether the isolation transformer was installed or not. This indicated that enough current was being supplied for the filaments. Of course, if there had actually been an H-K short, there would have been some slight smearing of the video due to the short.

I also checked the filament waveform from the HVT with a scope. It was a 28V pulse with a small dip in the center of the peak. With maximum brightness/contrast of the set, I could see no difference in either waveshape or amplitude, whether or not the isolation transformer was in-stalled—another indication of no excess loading on the HVT.

The point of the transformer, of course, is to allow the DC voltage of the filament to be the same as that of the cathode to which the filament is shorted, so as not to pull the cathode's voltage low. So if a filament pin is grounded on the CRT socket board, you may have to cut a strip out of the foil going to the grounded filament pin. Only the secondary of the isola-tion transformer should hook to the filament pins. The primary of the isolation transformer goes to the filament leads from the HVT, either the actual wires, or their connections on the CRT socket board. The trans-former is so light that it can be mounted by taping it to a low voltage wire from the CRT socket.

Additional Notes about the Homemade Isolation Transformer

I've been asked some questions about the homemade isolation transformer that I described in the article "A homemade isolation transformer to cure H-K shorts," which appeared in the September 1992 issue of *ES&T*. Read-ers asked questions like: How do you determine for sure that an isolation transformer is needed? How does it work? Exactly how do you hook it up? I hope that this article answers those questions. The original article is reprinted in its entirety after this article for easy reference to the informa-tion it contains.

Some Typical Filament Circuits

First, let's look at some typical examples of connections of the CRT fila-ment to the high voltage transformer (HVT). See *Figure 14-2*.

In *Figure 14-2A*, one lead of the HVT filament winding and one lead of the filament are grounded. In *Figure 14-2B*, one lead of the HVT fila-ment winding goes to ground through a resistor, and to B+ through an-other resistor. In *Figure 14-2C*, one lead of the HVT filament winding is

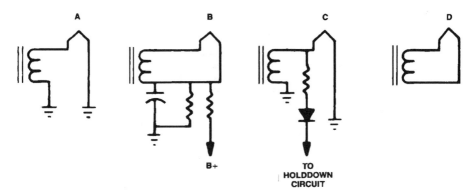

Figure 14-2. Some typical examples of CRT filament connections to the high voltage transformer (HVT). See text for details of each type of connection.

grounded and the other lead powers another circuit in the set. The point is that in each of these circuits there is a DC path from the filament to ground, or to another circuit.

But notice the circuit in *Figure 14-2D*. The transformer filament winding is connected only to the filament, with no grounds or connections to any other circuit. If all filament circuits were like the one in *Figure 14-2D*, then an isolation transformer would never be needed. The filament would already be isolated from any connections that would affect the picture if a heater-to-cathode (H-K) short occurred.

The terms "filament" and "heater," as used here, describe the same part of the CRT; the resistive component that heats the cathode and causes electron emission. The terms are interchangeable.

Why Isolating the Filament Removes the Symptom

By far the most important voltage that controls the conduction of the red, blue and green electron guns in the CRT is the voltage difference between the control grid (G1) and each cathode. For normal brightness, the cathodes are positive with respect to the control grid, as shown in Figure 2. Now let's assume that a short occurs between the red cathode and the filament, as shown by the heavy line in *Figure 14-3*.

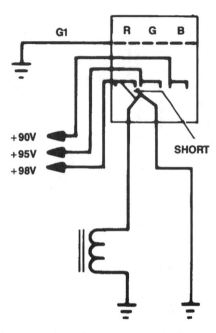

Figure 14-3. If a short develops between a cathode and its filament (heater), as shown by the heavy line in the drawing, it may cause picture problems, or possibly even shutdown of the set.

If the filament is grounded, then the short connects the red cathode to ground also, forcing the cathode voltage to drop to nearly zero volts. The red electron gun, with its cathode now grounded, will conduct heavily. This will cause a bright red screen. If the set has overcurrent protection, it may even shut down after the screen turns red, due to the heavy current through the CRT.

If the filament is not grounded, and is like the circuit of *Figure 14-2D*, neither the HVT filament winding nor the filament has any DC connections. If a cathode shorts to the filament, the DC voltage of the filament becomes the same as the dc voltage on the cathode which shorted to it. If the cathode voltage doesn't change, then the brightness of the shorted gun will not change. So even though the short remains, the picture is barely affected.

Ways to Isolate the Filament

So how can we make the other filament circuits in *Figure 14-2* as free of DC connections as the one in *Figure 14-2D*?

In *Figure 14-2A*, the ground wire of both the filament and the filament winding of the HVT could be removed from ground and wired together. Then the circuit would be the same as the circuit in *Figure 14-2D*. Naturally, this removes the symptom of the H-K short, as explained for *Figure 14-2D*.

The circuit in *Figure 14-2B* could be made like the one in *Figure 14-2D* by removing the capacitor and resistor to ground and the resistor to B+. Some circuits have a resistor in series with the filament. When making modifications, be sure to leave any series resistor in the circuit.

Though circuits like *Figure 14-2A* and *14-2B* can be modified to isolate them, it's often easier and quicker to install a separate isolation transformer than to modify the circuit.

In a circuit like *Figure 14-2A*, the HVT filament winding is sometimes physically grounded very close to the core of the transformer, so it's hard to splice and insulate a wire to the winding after it's removed from ground. If you decide to modify the circuit, be sure the result is like *Figure 14-2D* and that all splices are insulated.

Now we come to the circuit in *Figure 14-2C*. Here the transformer winding supplies another required voltage to the set. If you disconnect the ground, you will lose the other voltage source. The solution in this case is to use a separate isolation transformer to couple the HVT filament winding voltage to the filament while isolating the filament from ground.

The separate isolation transformer is simply a 1-to-1 ratio transformer which is connected into the circuit as shown in *Figure 14-4*. The secondary of the transformer has no DC connections to any other circuit, which

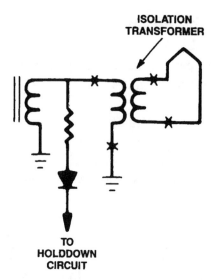

ISOLATION
TRANSFORMER

TO
HOLDDOWN
CIRCUIT

Figure 14-4. *The isolation transformer is connected into the circuit as shown here. This allows the filament to rise to the same DC voltage as the cathode. The presence of the isolation transformer eliminates the symptom of the H-K short, yet leaves the other connections of the HVT filament winding unchanged.*

allows the DC voltage on the filament to rise to the same DC voltage as the cathode to which the filament is shorted. Again, this eliminates the symptom of the H-K short, yet leaves the other connections of the HVT filament winding unchanged.

Confirming a Shorted CRT

Naturally, no amount of CRT heater isolation will help if the CRT is not the problem. Here's how I confirm whether or not the CRT has an H-K short.

If the screen constantly stays one bright color, I check the voltages on the CRT socket pins that go to the cathodes. If one cathode voltage is low, as I expect it will be, then I shut off the power and remove the socket from the CRT. Then I reapply the power and again check the voltage on the socket pin that had the low voltage. If that voltage is normal with the socket removed, then the CRT is shorted and it is time to try the isolation transformer.

If the voltage at the same pin of the socket is still low with the socket off the CRT, then the problem must be in the circuit supplying that cathode voltage, often the red, blue or green output transistor, or the IC supplying base drive to it.

If the symptom is intermittent, with the screen just briefly going bright red, green or blue, then it's harder to determine if the CRT is shorting. If I have the right adapter, I check the CRT with a CRT tester. If it shows a short, then I need check no further. If a short doesn't show on the CRT tester, I try tapping gently on the neck of the CRT with a screwdriver handle. If it still doesn't show a short, I remind myself that the problem is intermittent and that so far I haven't proved anything.

If I'm not able to check the CRT with a CRT tester, or I don't find a short using it, then I remove the socket from the CRT. When I find the socket pin of the cathode that would cause the bright screen color I've seen, I measure the voltage on that pin while I tap on the socket, the main board, and any connectors hooking the socket and the main board together.

If the reading drops, I know the problem is not the CRT. If the reading stays high, I assume the CRT is bad and install an isolation transformer. Then I let the set cook, occasionally tapping the circuit board and CRT socket. If the screen does not show the symptom in a couple of days (depending on how intermittent it was), then I conclude that it was a bad CRT.

If the symptom does reappear after the transformer is installed, then I know it is a circuit problem. But I don't remove the isolation transformer right away. A few times I've had an intermittent bright screen with both a circuit problem and a bad CRT. So I leave the transformer connected, and after I've repaired the circuit, I again let the set cook a couple of days. If the symptom is gone after the circuit repair, but then returns again after I take out the isolation transformer that I had installed, I know that both the circuit and the CRT were bad.

Installing the Isolation Transformer

The CRT filament in most sets made in the last twelve years or so is powered by the HVT, but if I have any doubt, I check the schematic or trace the filament wires to the HVT. The homemade isolation transformer won't work on a 60 Hz powered filament.

When I'm ready to install the transformer, I remove the socket from the CRT to avoid any possible damage to the CRT. I look carefully at each foil which goes to a filament pin on the CRT socket and follow one of these two procedures:

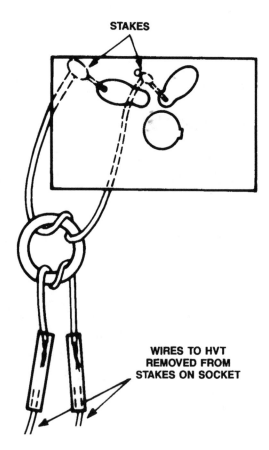

Figure 14-5. Connect the isolation transformer as shown here.

(a) If the only connections to each foil are a filament pin and a stake, with each stake having a single wire attached (in short, each conductor just ties the filament to the power lead), I remove the wire from each stake.

While the socket is removed and the wires are disconnected, I double-check with an ohmmeter to make sure that none of the foils have other connections. Then I slip lengths of heat shrink tubing over the wires from the HVT. I splice an end of one winding of the isolation transformer to each wire.

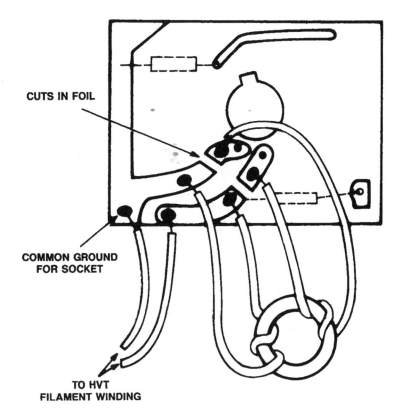

CUTS IN FOIL

COMMON GROUND FOR SOCKET

TO HVT FILAMENT WINDING

Figure 14-6. If either foil-to-a-filament pin has other connections, cut both foils so an area that goes to only the filament pin is on one side of each cut and hook the transformer as shown.

After I solder the splices, I use the heat shrink tubing to insulate each splice. Then I solder the ends of the other isolation transformer winding to the stakes to which the wires from the HVT were attached originally. See *Figure 14-5*.

(b) If either foil to a filament pin is long, or snakes around the board, it's almost a sure bet that it has other connections. If it does, I cut both foils so an area that goes to only the filament pin is on one side of each cut and hook the transformer, as shown in *Figure 14-6*.

This is probably the most universal way of installing the isolation transformer and should work on nearly all circuits. Before I install the isolation transformer as shown in *Figure 14-6*, I hold the core below the CRT board and estimate if the leads will be long enough for the core to hang below the CRT socket board.

If the leads will be too short, I wind another transformer with longer leads. I want the core to hang below the board where I can tape it to one of the wires from the filament winding of the HVT, or, if that's difficult, to another low-voltage wire-but definitely not to a focus or screen grid wire.

A General H-K Short Solution

As long as you keep in mind that the filament must have AC voltage across its terminals—but when there's an H-K short, its DC voltage must be the same as the voltage on the cathode to which it is shorted—you should be able to hook up the isolation transformer to any circuit variation you come across. At least until some engineer totally changes the game again.

Index

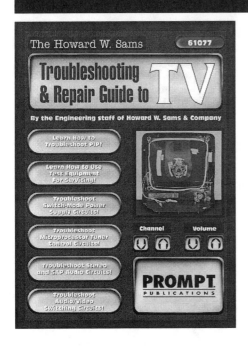

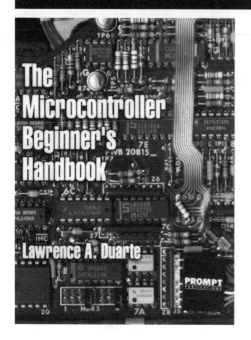
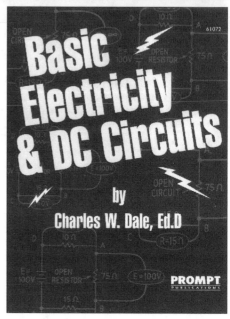

PROMPT
PUBLICATIONS

☞ **Dear Reader:** *We'd like your views on the books we publish.*

PROMPT® Publications, an imprint of Howard W. Sams & Company, is dedicated to bring-ing you timely and authoritative documentation and information you can use. You can help us in our continuing effort to meet you information needs. Please take a few moments to answer the questions below. Your answers will help us serve you better in the future.

1. **What is the title of the book you purchased?** _____
2. **Where do you usually buy books?** _____
3. **Where did you buy this book?** _____
4. **What did you like most about the book?** _____
5. **What did you like least?** _____
6. **Is there any other information you'd like included?** _____
7. **In what subject areas would you like us to publish more books?** (Please check the boxes next to your fields of interest.)

□ Audio Equipment Repair	□ Home Appliance Repair
□ Camcorder Repair	□ Mobile Communications
□ Computer Hardware	□ Security Systems
□ Electronic Concepts Theory	□ Sound System Installation
□ Electronic Projects/Hobbies	□ TV Repair
□ Electronic Reference	□ VCR Repair

8. **Are there other subjects that you'd like to see books about?** _____

9. **Comments** _____
_____
_____

Name _____
Address _____
City _____ **State/ZIP** _____
Online Address _____

Would you like a *FREE* PROMPT® Publications catalog? □Yes □No
Thank you for helping us make our books better for all of our readers. Please drop this postage-paid card into the nearest mailbox.

For more information about PROMPT® Publications, see your authorized Howard Sams distributor, or call 1-800-428-7267 for the name of your nearest PROMPT® Publications distributor.

An imprint of
Howard W. Sams & Company
A Bell Atlantic Company
2647 Waterfront Parkway, East Dr.
Suite 300
Indianapolis, IN 46214-2041

BUSINESS REPLY MAIL

FIRST CLASS MAIL   PERMIT NO. 1317   INDIANAPOLIS IN

POSTAGE WILL BE PAID BY ADDRESSEE

PROMPT PUBLICATIONS
AN IMPRINT OF HOWARD W SAMS & CO
2647 WATERFRONT PARKWAY EAST DR
SUITE 300
INDIANAPOLIS IN 46209-1418